SpringerBriefs in Computer Science

More information about this series at http://www.springer.com/series/10028

Giulia Traverso • Denise Demirel
Johannes Buchmann

Homomorphic Signature Schemes

A Survey

Giulia Traverso
Theoretische Informatik
Technische Universität Darmstadt
Darmstadt, Hessen, Germany

Denise Demirel
Theoretische Informatik
Technische Universität Darmstadt
Darmstadt, Hessen, Germany

Johannes Buchmann
Theoretische Informatik
Technische Universität Darmstadt
Darmstadt, Hessen, Germany

ISSN 2191-5768   ISSN 2191-5776   (electronic)
SpringerBriefs in Computer Science
ISBN 978-3-319-32114-1   ISBN 978-3-319-32115-8   (eBook)
DOI 10.1007/978-3-319-32115-8

Library of Congress Control Number: 2016935494

Printed on acid-free paper

This Springer imprint is published by Springer Nature
The registered company is Springer International Publishing AG Switzerland

Preface

In the last years, there has been an increasing interest in homomorphic signature schemes. Thus, many schemes have been proposed that are suitable for a lot of different applications. In this work, we overcome the extensive state of the art by presenting a survey of the existing approaches and the properties they provide. In addition, we look into three interesting use cases for homomorphic operations on authenticated data; these are electronic voting, smart grids, and electronic health records. We discuss their requirements, show to what extent the existing solutions meet these conditions, and highlight promising directions for future work.

Homomorphic signature schemes have been initially designed to establish authentication in network coding and to address pollution attacks (see [18]). However, since they allow for computations on authenticated data, they are also a useful primitive for many other applications. In fact, after Johnson et al. introduced a formal definition and a precise framework for homomorphic signatures in 2002 (see [46]), in the following years, many schemes have been presented and discussed. The first schemes proposed only allow to perform linear computations on authenticated data (e.g., [71, 72, 76], and [24]). These approaches have been further improved with respect to efficiency, security, and privacy [5–7, 15, 18, 21, 22, 22, 34, 36, 65]. In addition, to be more flexible, solutions have been developed supporting polynomial functions [14, 23, 43], or even coming without any restrictions on the functions themselves, so-called fully homomorphic signature schemes [19, 41]. However, all these solutions assume that each input signature has been generated using the same private key. To overcome this restriction, the homomorphic property has been added to the aggregate signature schemes [45, 74] allowing for operations on signatures generated using even different secret-public key pairs.

In this work, we start by providing a formal definition of these four types of homomorphic signature schemes. First, the passage from the digital signature schemes to the homomorphic ones is formally described, where the novelties introduced by the homomorphic property itself are highlighted. Afterward, it is described how to obtain the linearly homomorphic signature schemes from the merely homomorphic ones. And then, starting from the linearly homomorphic signature schemes, it is shown how to derive the schemes supporting polynomial functions and how to define the fully homomorphic signature schemes. Finally, schemes that allow computations on signatures generated using different secret-public key pairs are formally described.

Up to our knowledge, this survey is the first such work providing both a description of each single homomorphic signature scheme and a description of the whole general framework in a methodical and didactic approach. Indeed the survey proposed in [73] is not up to date, while in [20] the existing homomorphic signature schemes are just listed, without any deeper discussions. Furthermore, in this survey, we also discuss the possible use cases electronic voting, smart grids, and electronic health records. For each use case, concrete examples of how improvements can be achieved by the usage of homomorphic signature schemes are provided, together with the definition of the minimal requirements these schemes should fulfill. Furthermore, it is shown which of the currently existing homomorphic signature schemes are suitable for which of the use cases in question. When that is not the case, directions for future works are proposed.

In Chap. 1, the definition of general digital signature schemes is recalled, and the formal description of the homomorphic signature schemes is provided. Chapter 2 provides a description of the linearly homomorphic signature schemes, the homomorphic signature schemes for polynomial functions, the fully homomorphic signature schemes, and the homomorphic aggregate signature schemes. In Chap. 3, interesting properties of homomorphic signature schemes are discussed. The description of each of the currently existing homomorphic signature scheme and the properties they provide follow in Chap. 4. In Chap. 5, the usage of homomorphic signature schemes for each of the aforementioned use cases is presented. Finally, in Chap. 6, a conclusion is given and possible directions for future work are shown.

Darmstadt, Germany   Giulia Traverso
February 2016   Denise Demirel
                                                            Johannes Buchmann

Acknowledgments

This work has been co-funded by the European Union's Horizon 2020 research and innovation program under grant agreement no. 644962. In addition, it has received funding from the DFG as part of project Long-Term Secure Archiving Within the CRC 1119 CROSSING. We would like to thank also Lucas Shabhüser, Nina Bindel, Daniel Slamanig, and David Derler for the nice discussions.

Contents

Chapter 1
From Digital to Homomorphic Signature Schemes

Abstract A signature is a cryptographic primitive providing *integrity* and *authenticity*. Integrity means that the message signed has not been modified. Authenticity refers to the possibility of identifying its origin. In this chapter, first digital signature schemes are described and their security properties are discussed. Afterwards, the differences between those signature schemes and their homomorphic counterparts are highlighted and corresponding definitions are provided.

1.1 Digital Signatures

A digital signature scheme is defined over the following sets [40]:

- the messages space $\mathcal{M}$;
- the space of signed messages $\mathcal{Y}$;
- the set of secret keys $\mathcal{K}$;
- the set of public keys $\mathcal{K}'$.

In addition, the following three algorithms are defined:

- a probabilistic, polynomial-time algorithm *Set*: $1^{\lambda} \rightarrow \mathcal{K} \times \mathcal{K}'$ for the preliminary stage, where the secret key (used in the signing process) and the public key (used in the verification process) are chosen;
- a probabilistic, polynomial-time algorithm *Sig*: $\mathcal{K} \times \mathcal{M} \rightarrow \mathcal{Y}$ for signing;
- a deterministic algorithm *Vrf*: $\mathcal{K}' \times \mathcal{M} \times \mathcal{Y} \rightarrow \{0, 1\}$ for verification.

The algorithms *Set* and *Sig* are probabilistic. This means that some randomness is added to the input leading to a non-deterministic output. In addition those algorithms are polynomial-time bounded, as the number of steps, and consequently the running time, is bounded by a polynomial.

Now a definition of the role of each of the aforementioned algorithm is provided [56].

Definition 1.1. A *digital signature scheme* is a tuple of the following probabilistic, polynomial-time algorithms:

© The Author(s) 2016

G. Traverso et al., *Homomorphic Signature Schemes*, SpringerBriefs in Computer Science, DOI 10.1007/978-3-319-32115-8_1

- *Set*(1^λ). It takes as input a security parameter λ in unary. It outputs a secret key sk and the respective public key pk.
- *Sig*(sk, m). It takes as input a secret key sk and a message $m \in \mathcal{M}$. It outputs a signature $\sigma \in \mathcal{Y}$, which is the signature for the message m signed using the secret key sk.
- *Vrf*(pk, m, σ). It takes as input a public key pk, a message $m \in \mathcal{M}$, and a signature $\sigma \in \mathcal{Y}$. It outputs '1' if σ is a signature of the message m, signed using the secret key sk. It outputs '0' otherwise.

The verification algorithm is a fundamental step to check the authenticity of the signature. Specifically, if pk is the corresponding public key of the secret one sk, then the following equation must hold

$$\forall m \in \mathcal{M}, \quad \text{Vrf}(pk, m, \text{Sig}(sk, m)) = 1.$$

The verification is a public procedure. Indeed, *anyone*, using the public key, can verify that the signature belongs to the person who holds the secret key. Given a public key pk, a message-signature pair (m, σ) is *valid* if $Vrf(pk, m, \sigma) = 1$ [48].

Definition 1.2. A digital signature scheme is *correct* if for any message $m \in \mathcal{M}$ signed by the algorithm *Sig* and any secret key sk it holds that $Vrf(pk, m, Sig(sk, m)) = 1$.

Definition 1.1 formalizes what one would expect from a signature scheme in practice. There are two parties involved: a signer S and a receiver R. S computes a secret-public key pair (sk, pk). When it wants to communicate with R, S runs the signing algorithm for a message m and obtains a signature σ. Then it sends the pair (m, σ) to R. On its side, R verifies the authenticity of the message using the verification algorithm, i.e. checks that $Vrf(pk, m, Sig(sk, m)) = 1$, making sure that no errors or modification occurred during the transmission of m.

1.2   Digital Signature Schemes Security Definition

Besides the signer S and the receiver R, the presence of a possible attacker A is also to be taken into account. Such A has to be meant as a probabilistic, polynomial-time algorithm. Its goal is to create a *forgery* of a valid message-signature pair.

Definition 1.3. Let us assume that the signer S generates t legitimate message-signature pairs

$$(m_1, \sigma_1), (m_2, \sigma_2), \ldots, (m_t, \sigma_t)$$

using a certain digital signature scheme (*Set, Sig, Vrf*). The adversary A is able to forge it when it outputs a valid message-signature pair (m, σ), where m has not been previously signed by S. The pair (m, σ) is called an *existential forgery*.

Intuitively, we can say that a digital signature scheme is *secure* if no probabilistic, polynomial-time (opponent) algorithm A can output an existential forgery. That is, no adversary can get a valid (m, σ) in an efficient amount of time. In many applications, there has been lately the necessity to define a slightly different notion of existential forgery [48]. Assume a message-signature pair (m, σ) has been generated by S. In this case an adversary can perform a *strong existential forgery* if it is able to generate a valid message-signature pair (m, σ'), where $\sigma \neq \sigma'$.

Existential forgery and strong existential forgery are not the only possible types of forgery. However, as A is able to *select* the message to forge, they are the most sever results that an enemy A can end with after running the attack. In the framework of digital signature schemes, the strength and the power that A has depends also on the *control* it has over the messages that are signed. Three attacks can be distinguished. These are: *known-message attack*, *chosen-message attack*, and *adaptive chosen-message attack* [48]. In the following sections, each of them is separately discussed and formally defined.

1.2.1 Known-Message Attack

This is the scenario where the attacker A has no control over the messages that are signed by the legitimate signer [40]. However, it has knowledge of certain message-signature pairs. To model this situation we let the attacker randomly chose messages from $\mathcal{M}$ that are signed.

Definition 1.4 ([48]). A digital signature scheme (Set, Sig, Vrf) with security parameter λ is *existential unforgeable under known-message attack* if for all polynomials $t(\cdot)$ and probabilistic, polynomial-time adversaries A, the probability of success in the following game is negligible with respect to λ.

1) The messages $m_1, m_2, \ldots, m_t \in \mathcal{M}$ are chosen uniformly at random, for a certain $t := t(\lambda)$.
2) A key pair (sk, pk) is produced by the algorithm $Set(1^\lambda)$.
3) The message-signature pairs $(m_1, \sigma_1), (m_2, \sigma_2), \ldots, (m_t, \sigma_t)$ are computed. That is, for each message m_i, the algorithm $Sig(sk, m_i)$ is run to produce the respective signature σ_i.
4) The attacker A knows pk and all the pairs $(m_1, \sigma_1), (m_2, \sigma_2), \ldots, (m_t, \sigma_t)$, but not sk. A outputs a pair (m, σ)
5) A succeeds if the following two conditions hold:

   (a) $Vrf(pk, m, \sigma) = 1$;
   (b) $m \notin \{m_1, m_2, \ldots, m_t\}$.

Remark 1.1. The definition of digital signature schemes that are *strongly unforgeable under known message attacks* [48] is almost the same as for the weaker notion above. The only thing to take into account is the possibility that the forge might be on a previously legitimately signed message. Definition 1.4 is modified only regarding point 5).

5*) A succeeds if the following two conditions hold:

   (a) $Vrf(pk, m, \sigma) = 1$;
   (b) $(m, \sigma) \notin \{(m_1, \sigma_1), (m_2, \sigma_2), \ldots, (m_t, \sigma_t)\}$.

1.2.2 Chosen-Message Attack

The attacker A has some control over the messages that are signed by the legitimate signer. More precisely, A can select the sequences of messages that will be signed in the game, but this choice is made *a priori*, that is, before receiving any signature.

Definition 1.5 ([48]). A digital signature scheme (Set, Sig, Vrf) with security parameter λ is *existential unforgeable under chosen-message attack* if for all polynomials $t(\cdot)$ and probabilistic, polynomial-time adversaries A, the probability of success in the following game is negligible with respect to λ.

1) The messages $m_1, m_2, \ldots, m_t$ are chosen by A, for a certain $t := t(\lambda)$.
2) A key pair (sk, pk) is produced by the algorithm $Set(1^\lambda)$.
3) The message-signature pairs $(m_1, \sigma_1), (m_2, \sigma_2), \ldots, (m_t, \sigma_t)$ are computed. That is, for each message m_i, the algorithm $Sig(sk, m_i)$ is run to produce the respective signature σ_i.
4) The attacker A knows pk and all the pairs $(m_1, \sigma_1), (m_2, \sigma_2), \ldots, (m_t, \sigma_t)$, but not sk. A outputs a pair (m, σ)
5) A succeeds if the following two conditions hold:

   (a) $Vrf(pk, m, \sigma) = 1$;
   (b) $m \notin \{m_1, m_2, \ldots, m_t\}$.

Remark 1.2. The definition of digital signature schemes that are *strongly unforgeable under known-message attacks* [48] is almost the same as for the weaker notion above. As for the case of a known-message attack, the only thing to take into account is the possibility that the forge might be on a previous legitimately signed message. Definition 1.5 is modified only regarding point 5).

5*) A succeeds if the following two conditions hold:

   (a) $Vrf(pk, m, \sigma) = 1$;
   (b) $(m, \sigma) \notin \{(m_1, \sigma_1), (m_2, \sigma_2), \ldots, (m_t, \sigma_t)\}$.

Note that in the game described in Definition 1.5 the messages $m_1, m_2, \ldots, m_t$ are chosen before the attacker A gets to see the public key pk. It is possible to distinguish a further case, that is the game where the messages $m_1, m_2, \ldots, m_t$ are chosen after A gets access to the public key. The latter is called *directed chosen-message attack*, while the previous one described in Definition 1.5 is referred to as *generic chosen-message attack* [40].

1.2.3 Adaptive Chosen-Message Attack

Resilience against adaptive chosen-message attacks leads to the strongest notion of security that a digital signature algorithm can achieve. The attacker A has complete control over the messages that are signed. This comes from the fact that A can *choose adaptively*. Specifically, the attacker has the possibility to choose the message sequence after seeing the public key. Furthermore, A can select more messages to be signed based on the signatures received for the previously chosen messages.

More formally, the attacker has access to a "signing oracle" $\mathcal{O}_S(sk, \cdot)$. Such signing oracle uses the same secret key sk generated by the digital signature scheme itself to output a signature σ for any input message m the attacker wants.

Definition 1.6 ([48]). A digital signature scheme (Set, Sig, Vrf) with security parameter λ is *existential unforgeable under adaptive chosen-message attacks* if for all polynomials $t(\cdot)$ and probabilistic, polynomial-time adversaries A having access to a signing oracle $\mathcal{O}_S(sk, \cdot)$, the probability of success in the following game is negligible with respect to λ.

1) A key pair (sk, pk) is produced by the algorithm $Set(1^\lambda)$.
2) The attacker A knows pk and requests from the signing oracle $\mathcal{O}_S(sk, \cdot)$ signatures for as many messages as it likes and in an iteratively way. Let us denote with Q the sequence of messages queried by A from the signing oracle $\mathcal{O}_S(sk, \cdot)$.
3) A succeeds if it can output a message-signature pair (m, σ) for which the following two conditions hold:

   (a) $Vrf(pk, m, \sigma) = 1$;
   (b) $m \notin Q$.

It is common practice to use the notion presented in Definition 1.6 to describe the security of digital signature schemes. In particular, Definition 1.6 is often referred to as *EUF-CMA* and presented in the following form [40].

Definition 1.7 (EUF-CMA). A digital signature scheme (Set, Sig, Vrf) with security parameter λ is *existential unforgeable under adaptive chosen-message attacks* if for all polynomials $t(\cdot)$ and probabilistic, polynomial-time adversaries A having access to a signing oracle $\mathcal{O}_S(sk, \cdot)$, it holds that

$$Pr\left[(sk, pk) \leftarrow Set(1^\lambda), \ (m, \sigma) \leftarrow A^{\mathcal{O}_S(sk, \cdot)}(pk) : m \notin Q \wedge Vrf(pk, m, \sigma) = 1\right] = negl(\lambda).$$

For the sake of completeness, the definition of digital signature scheme that is *strongly unforgeable under adaptive chosen-message attacks* is also provided.

Definition 1.8 ([48]). A digital signature scheme (Set, Sig, Vrf) with security parameter λ is *strongly unforgeable under adaptive chosen-message attacks* if for all polynomials $t(\cdot)$ and probabilistic, polynomial-time adversaries A, the probability of success in the following game is negligible with respect to λ.

1) A key pair (sk, pk) is produced by the algorithm $Set(1^\lambda)$.
2) The attacker A knows pk and can request from the signing oracle $\mathcal{O}_S(sk, \cdot)$ signatures for as many messages as it likes and in an iteratively way. Let $Q^* = \{(m_i, \sigma_i)\}$, where m_i is the i-th query made by A to the signing oracle algorithm Sig and σ_i is the respective i-th response.
3) A succeeds if it can output a message-signature pair (m, σ) for which the following two conditions hold:

   (a) $Vrf(pk, m, \sigma) = 1$;
   (b) $m \notin Q^*$.

Remark 1.3. Another distinction about the power of an attacker A regards the *number* of forgeries it can generate. There are only two cases that are considered: either A gets an *unbounded* number of forgeries or just one. The first case is the one discussed so far in all the above definitions. The second case is called *one-time* signature. We refer to [48] for a formalization of the random-message attack, the known-message attack, and the adaptive chosen-message attack in the framework of one-time signature.

1.3 Homomorphic Signature Schemes

The work by Johnson et al. [46] is the first one providing a rigorous definition of homomorphic signature schemes. As the non-homomorphic ones, they are defined over a message space $\mathcal{M}$, a space of signed messages $\mathcal{Y}$, a space of secret keys $\mathcal{K}$, and a space of public keys $\mathcal{K}'$. In addition, homomorphic signature schemes provide a setup, signing, and verifying algorithm. The main difference between the non-homomorphic and homomorphic schemes is that the later ones support certain computations, i.e. the message space $\mathcal{M}$ and the space of signed messages $\mathcal{Y}$ are equipped with an operation.

In the following, two definitions of homomorphic signature schemes will be provided. The first one is the definition proposed by Johnson et al. in [46]. That is a very general one and basically it tells how the homomorphic property can be introduced to signature schemes. The second definition is more operative and provides a better description of the concrete construction of these schemes.

Definition 1.9. Assume we have a signature algorithm Sig and a verification algorithm Vrf, together with a binary operation '$\cdot$'. Then we say that Sig is a *homomorphic signature with respect to* · if it comes with an efficient family of binary operations $*_{pk} : \mathscr{Y} \times \mathscr{Y} \to \mathscr{Y}$ such that, having the messages $m, m' \in \mathscr{M}$ and the signatures $\sigma, \sigma' \in \mathscr{Y}$ for which

$$Vrf(pk, m, \sigma) = 1 = Vrf(pk, m', \sigma')$$

for a public key $k' \in \mathscr{K}'$, then there is a secret key $sk \in \mathscr{K}$ such that

$$\sigma *_{pk} \sigma' = Sig(sk, m \cdot m') \quad \text{and} \quad Vrf(pk, m \cdot m', \sigma *_{pk} \sigma') = 1.$$

With respect to the framework for digital signature schemes presented in Sect. 1.1, for homomorphic signature schemes there are some new elements to take into account:

- *the integer $N > 0$*: this value gives the maximum data size, i.e. the maximum number of messages operations can be performed on.
- *the set $\mathscr{F}$ of admissible functions*: this set defines which are the possible computations over the signed data that the homomorphic signature scheme can support. An element of $\mathscr{F}$ is a function $f : \mathscr{M}^N \to \mathscr{M}$.
- *the index i*: this index does not have an intrinsic meaning for the homomorphic signature scheme. It is just a practical tool to keep track of which of the messages of the data set $(m_1, m_2, \ldots, m_N) \in \mathscr{M}$ it has been working on.
- *the tag τ*: this element is necessary to define a strong notion of security also for homomorphic signature schemes.

In addition, a fourth algorithm is introduced. That is the algorithm $Eval : \mathscr{K}' \times \{0, 1\}^{\lambda} \times \mathscr{F} \times \mathscr{Y}^N \to \mathscr{Y}$, the core part of a homomorphic signature scheme. Indeed it allows anybody to compute on authenticated data, i.e. translating functions on messages into functions on signatures. In the following, these message-signature pairs will be denoted by $(m_1, \sigma_1), (m_2, \sigma_2), \ldots, (m_N, \sigma_N)$ and the tuple $(\sigma_1, \sigma_2, \ldots, \sigma_N)$ will be denoted by $\vec{\sigma}$.

Note that the other three algorithms behave slightly differently because of the new parameters they have in the input. Modeling on the operative definition for digital signatures and taking into account these differences, the corresponding homomorphic signatures are described as follows [34].

Definition 1.10. A *homomorphic signature scheme* is a tuple of the following probabilistic, polynomial-time algorithms:

- $Set(1^{\lambda}, N)$. It takes as input a security parameter λ in unary and an integer $N > 0$. It outputs a secret key sk and the respective public key pk. The public key determines the space of messages $\mathscr{M}$, the space of signatures $\mathscr{Y}$, and the set $\mathscr{F}$ of admissible functions $f : \mathscr{M}^N \to \mathscr{M}$.
- $Sig(sk, \tau, m, i)$. It takes as input a secret key sk, a tag $\tau \in \{0, 1\}^{\lambda}$, a message $m \in \mathscr{M}$, and an index $i \in \{1, 2, \ldots, N\}$. It outputs a signature $\sigma \in \mathscr{Y}$, computed using the secret key sk, which is the signature for the i-th message m of the data set tagged by τ.

- $Vrf(pk, \tau, m, \sigma, f)$. It takes as input a public key pk, a tag $\tau \in \{0, 1\}^\lambda$, a message $m \in \mathcal{M}$, a signature $\sigma \in \mathcal{Y}$, and a function $f \in \mathcal{F}$. It outputs '1' if σ is a valid signature for the message m. Such message m is the output of the function f over the data set tagged by τ, whose messages are signed using the public key pk. It outputs '0' otherwise.
- $Eval(pk, \tau, f, \overrightarrow{\sigma})$. It takes as input a public key pk, a tag $\tau \in \{0, 1\}^\lambda$, a function $f \in \mathcal{F}$ and a tuple of signatures $\overrightarrow{\sigma} \in \mathcal{Y}^N$. It outputs a signature $\sigma' \in \mathcal{Y}$, output of a function $f \in \mathcal{F}$ over the (tuple of) signatures $\overrightarrow{\sigma} \in \mathcal{Y}^N$. Such tuple $\overrightarrow{\sigma}$ corresponds to the signatures on the messages within the data set labeled by tag $\tau \in \{0, 1\}^\lambda$.

For homomorphic signature schemes *correctness* is defined as follows.

The index i that the algorithm *Sig* takes as input indicates that m is the i-th message in the list of messages $m_1, m_2, \ldots, m_N$. The projection can then be defined as

$$\pi_i : \mathcal{M}^N \to \mathcal{M} \quad \text{such that} \quad (m_1, m_2, \ldots, m_N) \mapsto m_i,$$

where $\pi_i \in \mathcal{F}$, $\forall\, i \in \{1, 2, \ldots, N\}$. A signature scheme is *correct* if the projection on any signed message $m \in \mathcal{M}$ and the output of $f(m_1, m_2, \ldots, m_N) \in \mathcal{M}$ leads to valid signatures. More precisely, this means that beyond verifying computations on multiple signatures, first of all the single signatures have to be valid by themselves. In the latter case, as denoted above, the verification algorithm takes as admissible function the projection on a single message (i.e. $Vrf(pk, \tau, m, \sigma, \pi_i)$), reducing Vrf to the usual verification algorithm of a digital signature (i.e. $Vrf(pk, \tau, m, \sigma)$). For this reason, it will be denoted with just four parameters as input, omitting the function π_i.

Definition 1.11. A homomorphic signature scheme is *correct* if for each output by the algorithm $Set(1^\lambda, N)$ the following conditions hold:

(1) For all $\tau \in \{0, 1\}^\lambda$, for all $m \in \mathcal{M}$, if σ is the output of $Sig(sk, \tau, m, i)$, then

$$Vrf(pk, \tau, m, \sigma) = 1.$$

(2) For all $\tau \in \{0, 1\}^\lambda$ and for all pairs (m_i, σ_i) for which $Vrf(pk, \tau, m_i, \sigma_i) = 1$, with $i \in \{1, 2, \ldots, N\}$,

$$Vrf(pk, \tau, f(\overrightarrow{m}), \overrightarrow{\sigma}, Eval(pk, \tau, f, \overrightarrow{\sigma})) = 1,$$

where $\overrightarrow{m} := (m_1, m_2, \ldots, m_N)$ and $\overrightarrow{\sigma} := (\sigma_1, \sigma_2, \ldots, \sigma_N)$.

1.4 Homomorphic Signature Schemes Security Definition

In the framework of digital signature schemes, being secure means that an attacker A cannot generate a new signature for a message without having the secret key. This security notion cannot be achieved by homomorphic signature schemes. That is due to their intrinsic design: as Johnson et al. pointed out in [46], no homomorphic signature will ever be safe against existential forgeries. Clearly, if the signature is homomorphic with respect to the operation '$*$', by definition, this means that from two signatures on the messages m_1 and m_2, one can directly compute a signature on $m_1 * m_2$ (without having access to the secret key). What can be required in this case is that no one is able to forge signatures on messages outside $\text{span}_*(m_1, m_2, \ldots, m_N)$.

Definition 1.12. Let us assume that the signer S generated N legitimate message-signature pairs

$$(m_1, \sigma_1), (m_2, \sigma_2) \ldots, (m_t, \sigma_N)$$

using a certain homomorphic signature scheme (*Set, Sig, Vrf, Eval*). The adversary A is able to forge it when it outputs a valid message-signature pair (m, σ), where m has not been previously signed by S and $m \notin \text{span}_*(m_1, m_2, \ldots, m_N)$. The pair (m, σ) is called an *existential forgery*.

Note that the notion of security for homomorphic signature schemes is correlated to one of the following two facts.

(1) The size of $\text{span}_*(m_1, m_2, \ldots, m_N)$ with respect to the number of elements $m_1, m_2, \ldots, m_N$ itself. What we wish to have is that a big set of messages span into a small space, or at least that a few messages do not have a large span.

(2) The hardness of decomposing a message m inside $\text{span}_*(m_1, m_2, \ldots, m_N)$. That is, the hardness of writing the message m in terms of $m_1, m_2, \ldots, m_N$.

If the "decomposition problem" is difficult, then the scheme can be secure even if the first requirement is not satisfied. The contrary also holds: if the scheme does not present a large span for a small set of messages, it is still secure even though the decomposition is easy. Since being able to quickly decompose a message may be useful also for the legitimate signer (see [57]), we do not consider the issue of having a hard decomposition. Therefore, the definition of security of a homomorphic signature is presented as follows (see [46]).

The *advantage* of an adversary A is defined as the probability that it outputs a valid signature (m, σ) for a message $m \notin \text{span}_*(m_1, m_2, \ldots, m_N)$, after queries on messages $m_1, m_2, \ldots, m_N$.

Definition 1.13. A homomorphic signature scheme (*Set, Sig, Vrf, Eval*) is (t, q, ε)-*secure* against existential forgeries (with respect to $*$), if every probabilistic, polynomial-time adversary A making no more than q chosen-message queries to the algorithm Sig and running in time at most t has advantage at most ε:

$$\text{Adv } A \leq \varepsilon,$$

where $\varepsilon > 0$ is a real number.

Definition 1.13 does not consider the attacker model discussed in Sect. 1.4. Note that in the framework of homomorphic signature schemes it is not straightforward that "secure" means resilient against adaptive chosen-messages attacks. Indeed there are homomorphic signature schemes where the attacker A cannot ask for queries adaptively after seeing some message-signature pairs of its choice. This subject will be deeply discussed in Sect. 3.3, where a more operative definition based on games is provided.

Chapter 2
Homomorphic Signature Schemes

Abstract In this chapter two types of signature schemes satisfying homomorphic properties are presented. In the first section a description of the homomorphic signature schemes suitable in the single-user scenario is provided. In the second section the homomorphic signature schemes that support the multi-user case are presented.

2.1 Homomorphic Signature Schemes for the Single-User Scenario

There are three different types of homomorphic signature schemes for the single-user case. In fact, the whole set of homomorphic signatures can be divided according to the admissible functions each scheme supports. Specifically, we can distinguish schemes providing:

- linear functions, so called linearly homomorphic signature schemes;
- polynomial functions, so called homomorphic signature schemes for polynomial functions;
- arbitrary functions, so called fully homomorphic signature schemes.

The signatures are presented in the same order as listed above. That is, they are discussed with respect to a set of possible admissible functions which is less and less restrictive. For each section, the differences with respect to the general definition of homomorphic signatures are highlighted and the evolution from linearly up to fully homomorphic signature schemes is shown.

2.1.1 Linearly Homomorphic Signature Schemes

A linear homomorphic signature scheme [45] allows to perform linear functions over signed messages. With respect to the general definition of homomorphic signature schemes, there are some differences to point out.

© The Author(s) 2016
G. Traverso et al., *Homomorphic Signature Schemes*, SpringerBriefs in Computer Science, DOI 10.1007/978-3-319-32115-8_2

- The messages space $\mathcal{M}$ is the vector space $\mathbb{F}_p^N$ of dimension N defined over the finite field $\mathbb{F}_p$, for a prime number p. Such p is an additional output of the setup algorithm *Set*.
- The messages are vectors. More precisely, they are elements $v \in \mathbb{F}_p^N$, i.e. $v = (a_1, a_2, \ldots, a_N)$ where $a_i \in \mathbb{F}_p$.
- If we consider the vectors $v_1, v_2, \ldots, v_N$, then the set $\mathcal{F}$ of admissible functions $f \in \mathcal{F}$ are all possible linear combinations in the $\mathbb{F}_p$-linear span of $v_1, v_2, \ldots, v_N$.

The homomorphic property in this context is specified as follows. Given *one* signature *per* message $v_1, v_2, \ldots, v_N$ in $\mathbb{F}_p^N$, *anyone* can compute a signature for a vector $v' \in \mathbb{F}_p^N$, where:

- $v' := f(\vec{v}) = \sum_{i=1}^N c_i v_i$, for $\vec{v} := (v_1, v_2, \ldots, v_N)$ and
- $c_1, c_2, \ldots, c_N \in \mathbb{F}_p$.

The definition of linearly homomorphic signatures follows [18].

Definition 2.1. A *linearly homomorphic signature scheme* is a tuple of the following probabilistic, polynomial-time algorithms:

- *Set*$(1^\lambda, N)$. It takes as input a security parameter λ in unary and an integer $N > 0$. It outputs a secret key *sk*, the respective public key *pk*, and a prime number p. The public key determines the space of messages $\mathbb{F}_p^N$, the space of signatures $\mathbb{F}_p^N$, and the set $\mathcal{F}$ of admissible functions $f : \mathbb{F}_p^N \to \mathbb{F}_p^N$.
- *Sig*(sk, τ, v, i). It takes as input a secret key *sk*, a tag $\tau \in \{0, 1\}^\lambda$, a vector $v \in \mathbb{F}_p^N$, and an index $i \in \{1, 2, \ldots, N\}$. It outputs a signature $\sigma \in \mathbb{F}_p^N$, computed using the secret key *sk*, which is the signature for the i-th message v of the data set tagged by τ.
- *Vrf*(pk, τ, v, σ, f). It takes as input a public key *pk*, a tag $\tau \in \{0, 1\}^\lambda$, a vector $v \in \mathbb{F}_p^N$, a signature $\sigma \in \mathbb{F}_p^N$, and a function $f \in \mathcal{F}$. It outputs '1' if σ is a valid signature for the vector v. Such vector v is output of the function f over the data set tagged by τ, whose messages are signed using the public key *pk*. It outputs '0' otherwise.
- *Eval*$(pk, \tau, f, \vec{\sigma})$. It takes as input a public key *pk*, a tag $\tau \in \{0, 1\}^\lambda$, a function $f \in \mathcal{F}$, and a tuple of signatures $\vec{\sigma} \in \mathbb{F}_p^N$. It outputs a signature $\sigma' = \sum_{i=1}^N c_i \sigma_i \in \mathbb{F}_p^N$ for a function $f \in \mathcal{F}$ over the (tuple of) signatures $\vec{\sigma} \in \mathbb{F}_p^N$. Such tuple $\vec{\sigma}$ corresponds to the signatures on the messages within the data set labeled by tag $\tau \in \{0, 1\}^\lambda$.

For the *correctness*, we refer to Definition 1.11, where the message $m \in \mathcal{M}$ is a vector $v \in \mathbb{F}_p^N$ and $f(\vec{m}) = \sum_{i=1}^N c_i v_i$.

2.1.2 Homomorphic Signature Schemes for Polynomial Functions

A homomorphic signature scheme for polynomial functions is a signature scheme that allows to compute polynomial functions over signed messages. The first of such schemes is proposed in [14], where the polynomials are multivariates of bounded degree. It can be seen as a generalization of linearly homomorphic schemes, since in the linearly case the set of admissible functions is a polynomial of degree one. The following definition of homomorphic signatures for polynomial functions is a generalization of the original one presented in [14].

As described in [14], the general framework of such signatures is composed of the following elements.

- The message space is a finite field $\mathbb{F}_p$, for a prime number p.
- The space of signed messages $\mathscr{Y}$ is the polynomial ring $R := \mathbb{Z}[x]/\langle F(x)\rangle$, for a monic irreducible polynomial $F(x) \in \mathbb{Z}[x]$ of degree d. Such polynomial is the new output of the algorithm *Set*.
- The set of admissible functions $\mathscr{F} \subset \mathbb{F}_p[x_1, \ldots, x_N]$ for the variables $x_1, \ldots, x_N$, with coefficients in $\{-y, \ldots, y\}$ and degree at most d, where y and d are positive integers.

The homomorphic property for this type of signature schemes follows. Given *one* signature *per* message $m_1, m_2, \ldots, m_N$, anyone can compute a signature for the polynomial $f(\overrightarrow{m}) = \sum_{j=1}^{\ell} c_j Y_j(\overrightarrow{m})$ where:

- $\overrightarrow{m} = (m_1, m_2, \ldots, m_N)$;
- $\ell := \binom{N+d}{d} - 1$;
- $\{Y_j\}_{j=1}^{\ell}$ is a set of non-constant monomials $x_1^{e_1}, x_2^{e_2} \ldots, x_N^{e_N}$ of degree $\sum e_N \leq d$;
- $c_1, c_2, \ldots, c_l$ are coefficients in $\mathbb{F}_p$.

Definition 2.2. A *homomorphic signature scheme for polynomial functions* is a tuple of the following probabilistic, polynomial-time algorithms:

- *Set*$(1^{\lambda}, N)$. It takes as input a security parameter λ in unary and an integer $N > 0$. It outputs a secret key sk, the respective public key pk, a prime number p, and a monic irreducible polynomial $F(x) \in \mathbb{Z}[x]$ of degree d. The public key determines the space of messages $\mathbb{F}_p$, the space of signatures R, and the set $\mathscr{F} \subset \mathbb{F}_p[x_1, \ldots, x_N]$ of admissible functions, where $y = poly(\lambda)$ and $d = \mathscr{O}(1)$.
- *Sig*(sk, τ, m, i). It takes as input a secret key sk, a tag $\tau \in \{0, 1\}^{\lambda}$, a message $m \in \mathbb{F}_p$, and an index $i \in \{1, 2, \ldots, N\}$. It outputs a signature $\sigma \in R$, computed using the secret key sk, which is the signature for the i-th message m of the data set tagged by τ.
- *Vrf*(pk, τ, m, σ, f). It takes as input a public key pk, a tag $\tau \in \{0, 1\}^{\lambda}$, a message $m \in \mathbb{F}_p$, a signature $\sigma \in R$, and a function $f \in \mathscr{F}$. It outputs '1' if σ is the valid

signature for the message m. Such message m is the output of the function f over the data set tagged by τ, whose messages are signed using the public key pk. It outputs '0' otherwise.

- $Eval(pk, \tau, f, \overrightarrow{\sigma})$. It takes as input a public key pk, a tag $\tau \in \{0, 1\}^{\lambda}$, a function $f \in \mathscr{F}$, a tuple of signatures $\overrightarrow{\sigma} \in R$. It outputs a signature $\sigma' = f(\overrightarrow{\sigma}) \in R$ for a function $f \in \mathscr{F}$ over the (tuple of) signatures $\overrightarrow{\sigma} \in R$. Such tuple $\overrightarrow{\sigma}$ corresponds to the signatures on the messages within the data set labeled by tag $\tau \in \{0, 1\}^{\lambda}$.

For the *correctness*, we refer to Definition 1.11, where $f(\overrightarrow{m}) = \sum_{j=1}^{\ell} c_j Y_j(\overrightarrow{m})$.

Remark 2.1. Note that in [14], the computation of $f \in \mathbb{F}_p[x_1, x_2, \dots, x_N]$ over the tuple of signatures $\overrightarrow{\sigma}$ is actually performed in two steps. First, f is lifted to a function, $\hat{f} \in \mathbb{Z}[x_1, x_2, \dots, x_N]$ defined as $\hat{f} := \sum_{j=1}^{l} c_j Y_j(x_1, x_2, \dots, x_N)$, where $c_1, c_2, \dots, c_N$ are integer coefficients. Then, σ' is computed as the output of $\hat{f}(\overrightarrow{\sigma})$.

2.1.3 Fully Homomorphic Signatures

Using fully homomorphic signature schemes there are no restrictions with respect to the operations that can be performed on the signed messages. Now, being allowed to use both $+$ and $\times$ over a field $\mathbb{F}_p$, it is possible to evaluate any function. Such function is now described by a *circuit* C with a certain size and a certain depth d. We do not propose here the definition of a fully homomorphic signature scheme, since it is almost the same as for a general homomorphic signature (Definition 1.10). However, the few variations to take into account are discussed.

- The function is seen as a circuit, which is denoted as $C : \mathscr{M}^N \to \mathscr{M}$.
- Instead of the set of admissible functions, a circuit family $\mathscr{C}$ is employed.
- The algorithm *Setup* outputs the secret key and the public key, but not the set of admissible functions anymore.
- The notion of correctness remains the same with the remark that the circuit C can also be a projection circuit P_i, i.e. $P(m_1, \dots, m_N) = m_i$. This means that the correctness must also hold for single-message signatures (see [19]).

For the *correctness*, due to the generality of the function that a fully homomorphic scheme supports, we refer directly to Definition 1.11, where the description of such f covers all types of functions.

2.2 Homomorphic Signature Schemes for the Multi-Users Scenario

The homomorphic signature schemes presented so far are suitable for the single-user scenario. There is only one signer owning the secret key that generates authenticated messages. When there are multiple users involved in the signing process, then homomorphic signature schemes supporting the multi-users case are needed. In this scenario we want to perform operations on signatures on messages signed by different users, each of them with its own private key. Among these schemes, the following two types of multi-users signature schemes can be distinguished:

- multi sources homomorphic signature schemes;
- homomorphic aggregate signature schemes.

2.2.1 Multiple Sources Homomorphic Signature Schemes

As for the linearly homomorphic signature schemes, the multiple sources homomorphic schemes were at first introduced by Agrawal et al. in [3] to address pollution attacks within the network coding framework (for a precise explanation we refer to [18]). Multi sources network coding refers to the situation where several sources transmit data, instead of a single one. For this reason in literature the latter schemes are called "multiple sources network coding signature schemes". Though, we refer to them omitting the link to network coding, as such signature schemes allow for supporting multiple users, and not only multiple sources.

As the homomorphic signature schemes for a single user, the multiple sources homomorphic ones are defined over a message space $\mathcal{M}$, a space of signed messages $\mathcal{Y}$, both of them equipped with an operation, a space of secret keys $\mathcal{K}$, and a space of public keys $\mathcal{K}'$. They still provide a setup, signing, verifying, and evaluating algorithm. All of them are probabilistic, polynomial-time algorithms, except for the verification algorithm that is deterministic. Though, in order to support multiple users, some differences are introduced in the definition of multiple sources homomorphic signature schemes.

- The algorithm *Set* takes as input a security parameter and a positive integer N, which stands for the maximum number of users the scheme can support. In addition, the algorithm *Set* returns N secret-public key pairs.
- The algorithm *Sig* takes as input a secret key, a message, and an index $i \in \{1, 2, \ldots, N\}$, in order to specify the user i together with its secret key sk_i and message m_i
- The algorithm *Vrf* takes as input a string of N public keys $pk_1, pk_2, \ldots, pk_N$ (indicated by $\vec{pk}$), one for each signer, a tag, a message, a signature, and an admissible function.
- The algorithm *Eval* takes as input a string of N public keys $\vec{pk}$, a tag, an admissible function, and a string of N signatures $\sigma_1, \sigma_2, \ldots, \sigma_N$ (indicated by $\vec{\sigma}$), one for each signer.

In the following we provide a more precise definition of multiple sources homomorphic signature schemes. The original definition provided by Agrawal et al.

in [3] is specific for multiple sources network coding. We adapt the definition to the more general multi-users scenario.

Definition 2.3. A *multiple sources homomorphic signature scheme* is a tuple of the following probabilistic, polynomial-time algorithms:

- $Set(1^\lambda, N)$. It takes as input a security parameter λ in unary and an integer $N > 0$. It outputs N pairs (sk_i, pk_i) of secret and public keys, one for each user i. The public key determines the space of messages $\mathcal{M}$, the space of signatures $\mathcal{Y}$, and the set $\mathcal{F}$ of admissible functions $f : \mathcal{M}^N \rightarrow \mathcal{M}$.
- $Sig(sk, \tau, m, i)$. It takes as input a secret key sk, a tag $\tau \in \{0, 1\}^\lambda$, a message $m \in \mathcal{M}$, and an index $i \in \{1, 2, \ldots, N\}$. It outputs a signature $\sigma \in \mathcal{Y}$, computed using the i-th secret key sk, which is the signature for the i-th message m of the data set tagged by τ.
- $Vrf(\overrightarrow{pk}, \tau, m, \sigma, f)$. It takes as input a public keys' string $\overrightarrow{pk}$, a tag $\tau \in \{0, 1\}^\lambda$, a message $m \in \mathcal{M}$, a signature $\sigma \in \mathcal{Y}$, and a function $f \in \mathcal{F}$. It outputs '1' if σ is a valid signature for the message m. Such message m is the output of the function f over the data set tagged by τ, whose messages are signed using the public keys in the string $\overrightarrow{pk}$. It outputs '0' otherwise.
- $Eval(\overrightarrow{pk}, \tau, f, \overrightarrow{\sigma})$. It takes as input a public keys' string $\overrightarrow{pk}$, a tag $\tau \in \{0, 1\}^\lambda$, a function $f \in \mathcal{F}$ and a tuple of signatures $\overrightarrow{\sigma} \in \mathcal{Y}^N$. It outputs a signature $\sigma' \in \mathcal{Y}$ on the output of a function $f \in \mathcal{F}$ over the (tuple of) signatures $\overrightarrow{\sigma} \in \mathcal{Y}^N$. Such tuple $\overrightarrow{\sigma}$ corresponds to the signatures on the messages within the data set labeled by tag $\tau \in \{0, 1\}^\lambda$.

The definition *correctness* takes into account both the homomorphic property and the fact that multiple public keys may be involved in the verification process. When a single signature has to be verified the algorithm *Vrf* takes as input one public key only. As for the homomorphic signature schemes in the single-user case, the admissible function f is meant as a projection from the data set to the message in question and will not be specified.

Definition 2.4. A homomorphic aggregate signature is *correct* if for each of the N secret-public key pair (sk, pk) output of the algorithm $Set(1^\lambda, N)$, the following conditions are valid:

(1) For all $\tau \in \{0, 1\}^\lambda$ and $i \in \{1, 2, \ldots, N\}$, if σ is the output of $Sig(sk, \tau, m, i)$, then $Vrf(pk, \tau, m, \sigma) = 1$.
(2) For all $\tau \in \{0, 1\}^\lambda$ and $i \in \{1, 2, \ldots, N\}$, for all pairs (m_i, σ_i) and (sk_i, pk_i) with $Vrf(pk_i, \tau, m_i, \sigma_i) = 1$

$$Vrf(\overrightarrow{pk}, \tau, f(\overrightarrow{m}), \overrightarrow{\sigma}, Eval(\overrightarrow{pk}, \tau, f, \overrightarrow{\sigma})) = 1,$$

where $\overrightarrow{m} := (m_1, m_2, \ldots, m_N)$ and $\overrightarrow{\sigma} := (\sigma_1, \sigma_2, \ldots, \sigma_N)$.

2.2.1.1 Multiple Sources Linearly Homomorphic Signature Schemes

The existing multiple sources homomorphic signature schemes proposed so far support linear operations over signed messages. We refer to these schemes as *multiple sources linearly homomorphic signature schemes*, even if in literature the adjective "linearly" is omitted, like for the aggregate case.

The definition of multiple sources linearly homomorphic signature schemes is quite similar to the single-user counterpart and can be easily derived from Definition 2.3. In the following, the changes to take into account to allow a multiple sources homomorphic signature scheme to support linear combinations on signatures are described. For a more formal definition we refer to [75].

- The messages space $\mathcal{M}$ and the signatures space $\mathcal{Y}$ are the vector space $\mathbb{F}_p^N$ of dimension N defined over the finite field $\mathbb{F}_p$, for a prime number p. Such p is the new output of the algorithm *Set*.
- The messages are vectors $v \in \mathbb{F}_p^N$, i.e. $v = (a_1, a_2, \ldots, a_N)$ where $a_i \in \mathbb{F}_p$.
- The set of the admissible functions $f \in \mathcal{F}$ are all the possible linear combinations in the $\mathbb{F}_p$-linear span of $v_1, v_2, \ldots, v_N$.

2.2.2 Homomorphic Aggregate Signature Schemes

In this section we discuss homomorphic aggregate signature schemes. Aggregate signatures allow to aggregate different signatures on different messages, signed by different users, each of them with its own secret key. This type of signature schemes has been firstly introduced by Boneh et al. in [17]. In the case we want to perform operations on the signatures rather then only aggregating them we need so-called *homomorphic aggregate signature schemes*. In the following, we first discuss the definition and the framework of aggregate signatures are. Then we define homomorphic aggregate signatures and finally describe the schemes supporting linear functions.

2.2.2.1 Aggregate Signature Schemes

An aggregate signature scheme combines multiple signatures into a single one. Assume we have N different messages and their N respective signatures generated by different secret-public key pairs. If the signatures have been generated using an *aggregate signature scheme* it is possible to generate a single signature for all the N messages.

More precisely, suppose that N users $u_1, u_2, \ldots, u_N$ want to obtain a signature σ which is an aggregation of the respective signatures $\sigma_1, \sigma_2, \ldots, \sigma_N$. Assume in addition that each message-signature pair (m_i, σ_i) has been generated by user u_i, using its own secret-public key pair (sk_i, pk_i). Then it is possible to aggregate these

N signatures into a single one. This signature should be short, i.e. it should not be longer than the original N signatures. Another important property of such scheme is to be *incremental*. That is, after aggregating N signatures $\sigma_1, \sigma_2, \ldots, \sigma_N$ and receiving the signature σ, it is always possible to aggregate σ and a further one σ_{N+1}. This means that we do not have to start the process from the scratch and run another aggregate signature scheme for $\sigma_1, \sigma_2, \ldots, \sigma_N, \sigma_{N+1}$. Instead, it is possible to generate the final signature σ' out of σ_{N+1} and σ.

As for any other signature, also the aggregate one is defined over the messages space $\mathcal{M}$, the signatures space $\mathcal{Y}$, and the set of secret and public keys $\mathcal{K}$ and $\mathcal{K}'$, respectively. Furthermore, it defines the standard algorithms *Set*, *Sig*, and *Vrf*.

- The algorithm *Set* chooses the secret key (used in the signing process) and the respective public key (used in the verification process) for each user.
- The algorithm *Sig* takes as input a secret key, a message, and the index $i \in \{1, 2, \ldots, N\}$, of user u_i and outputs a signature.
- The algorithm *Vrf* takes as input a message, a signature, and a string of N public keys $pk_1, pk_2, \ldots, pk'_N$ (indicated by $\overrightarrow{pk}$), one for each signer and checks the correctness of the signature.

In addition a new algorithm Agg_σ is introduced.

- The algorithm Agg_σ takes as input the signatures $\sigma_1, \sigma_2, \ldots, \sigma_N$, aggregate them, and outputs the resulting signature σ.

More precise, aggregate signature schemes are defined as follows.

Definition 2.5. An *aggregate signature scheme* is a tuple of the following probabilistic, polynomial-time algorithms:

- $Set(1^\lambda)$. It takes as input a security parameter λ in unary. It outputs a secret-public key pair (sk_i, pk_i) for each user i. The public keys determine the space of messages $\mathcal{M}$ and the space of signatures $\mathcal{Y}$.
- $Sig(sk, m, i)$. It takes as input a secret key sk, a message $m \in \mathcal{M}$, and an index $i \in \{1, 2, \ldots, N\}$. It outputs a signature $\sigma \in \mathcal{Y}$, which is the signature for the i-th message m, by means of the i-th secret key sk.
- $Vrf(\overrightarrow{pk}, m, \sigma)$. It takes as input the public keys' string $\overrightarrow{pk}$, a message $m \in \mathcal{M}$, and a signature $\sigma \in \mathcal{Y}$. It outputs '1' if σ is a valid signature for the message m, signed using the public keys $\overrightarrow{pk}$. It outputs '0' otherwise.
- $Agg_\sigma(\overrightarrow{pk}, \overrightarrow{m}, \overrightarrow{\sigma})$. It takes as input a public keys' string $\overrightarrow{pk}$, a messages's string $\overrightarrow{m} \in \mathcal{M}$, and a signatures' string $\overrightarrow{\sigma} \in \mathcal{Y}$. It outputs a signature $\sigma_{agg} \in \mathcal{Y}$, which is the aggregate signature of the signatures in $\overrightarrow{\sigma}$ of the messages in $\overrightarrow{m}$, signed using the public keys in $\overrightarrow{pk}$, respectively.

Now we give the definition of *correctness*. Roughly speaking, an aggregate signature scheme is correct if the verification holds for the independent signatures over the single messages and for the signature over the aggregated message. In the first case, the algorithm *Vrf* takes as input just one public key pk, that is the one corresponding to the secret key sk by which the single message has been signed. Therefore in this situation the algorithm Vrf coincides to the usual one defined for a classical digital signature scheme.

Definition 2.6. An aggregate signature scheme is *correct* if for each output (sk, pk) of the algorithm $Set(1^\lambda)$, the following conditions are valid:

(1) For all $i \in \{1, 2, \ldots, N\}$, if σ is the output of $Sig(sk, m, i)$, then $Vrf(pk, m, \sigma) = 1$.
(2) For all $i \in \{1, 2, \ldots, N\}$, if σ_i is the output of $Sig(sk, \tau, m, i)$, then $Vrf(\overrightarrow{pk}, \overrightarrow{m}, Agg_\sigma(\overrightarrow{pk}, \overrightarrow{m}, \overrightarrow{\sigma})) = 1$.

Remark 2.2. Everything described so far also holds for any arbitrary subset U of the N users, where $0 < |U| < N$.

2.2.2.2   Homomorphic Aggregate Signature Schemes

Homomorphic aggregate signature schemes combine two properties [74]. They are at the same time:

- a signature scheme that aggregates signatures produced by different users (*aggregate signature schemes*);
- a signature scheme that performs computations on signatures using an admissible function (*homomorphic signature schemes*).

In the following, the homomorphic property is added (that is, the possibility to compute on authenticated data) to the definition of aggregate signature schemes. Some differences have to be taken into account.

- The algorithm Set takes as input a security parameter and, in addition, an integer N, which stands for the maximum number of users the scheme can support. Therefore, the parameter N has to be decided a priori and this means that the incremental property of aggregate signatures is lost.
- The algorithm Agg_σ takes as input a public keys' string, a messages' string, and a signatures' string. In addition, it takes as input a tag $\tau \in \{0, 1\}^\lambda$ and an admissible function $f \in \mathscr{F}$. Instead of simply aggregating the signatures also function f is applied.
- A new algorithm is introduced, that is the algorithm Agg_m. It takes as input a public keys' string, a tag $\tau \in \{0, 1\}^\lambda$, a signatures' string, and an admissible function $f \in \mathscr{F}$. It performs f on the messages $m_1, m_2, \ldots, m_N$.
- The algorithm Sig takes as input a secret key, a message, and an index. In addition, it takes as input a tag $\tau \in \{0, 1\}^\lambda$.
- The algorithm Vrf takes as input a public keys' string, and a message. In addition, it takes as input the admissible function $f \in \mathscr{F}$.

The following definition formally addresses the modifications discussed above [74].

Definition 2.7. A *homomorphic aggregate signature scheme* is a tuple of the following probabilistic, polynomial-time algorithms:

- $Set(1^\lambda, N)$. It takes as input a security parameter λ in unary and an integer $N > 0$. It outputs N pairs (sk_i, pk_i) of secret and public keys, one for each user i. The public keys determine the space of messages $\mathcal{M}$, the space of signatures $\mathcal{Y}$, and the set $\mathcal{F}$ of admissible functions $f : \mathcal{M}^N \to \mathcal{M}$.
- $Sig(sk, \tau, m, i)$. It takes as input a secret key sk, a tag $\tau \in \{0,1\}^\lambda$, a message $m \in \mathcal{M}$, and an index $i \in \{1, 2, \ldots, N\}$. It outputs a signature $\sigma \in \mathcal{Y}$, computed using the i-th secret key sk, which is the signature for the i-th message m of the data set tagged by τ.
- $Vrf(\overrightarrow{pk}, \tau, m, \sigma, f)$. It takes as input a public keys' string $\overrightarrow{pk}$, a tag $\tau \in \{0,1\}^\lambda$, a message $m \in \mathcal{M}$, a signature $\sigma \in \mathcal{Y}$, and an admissible function $f \in \mathcal{F}$. It outputs '1' if σ is a valid signature for the message m, signed using the public keys $\overrightarrow{pk}$, output of the function f over the data set tagged by τ. It outputs '0' otherwise.
- $Agg_m(\overrightarrow{pk}, \tau, \overrightarrow{m}, f)$. It takes as input a public keys' string $\overrightarrow{pk}$, a tag $\tau \in \{0,1\}^\lambda$, a messages' string $\overrightarrow{m} \in \mathcal{M}$, and an admissible function $f \in \mathcal{F}$. It outputs an aggregated message $m_{Agg} \in \mathcal{M}$ by applying function f on the messages $\overrightarrow{m}$ in the data set labeled by tag τ, coming from the users with the public keys $\overrightarrow{pk}$.
- $Agg_\sigma(\overrightarrow{pk}, \tau, \overrightarrow{\sigma}, f)$. It takes as input a public keys' string $\overrightarrow{pk}$, a tag $\tau \in \{0,1\}^\lambda$, a signatures' string $\overrightarrow{\sigma} \in \mathcal{M}$, and an admissible function $f \in \mathcal{F}$. It outputs an aggregated signature σ_{Agg} by applying function f on the signatures $\overrightarrow{\sigma}$. The signatures in the string $\overrightarrow{\sigma}$ are the signatures on the messages within the data set labeled by tag τ, coming from the users with the public keys $\overrightarrow{pk}$.

The *correctness* definition takes into account the new algorithm Agg_m and the introduction of the homomorphic property. When a single signature is verified, as for the aggregate signature schemes, the algorithm Vrf takes as input just one public key. In addition, as for the homomorphic signature schemes, the admissible function f is meant as a projection from the data set to the message in question.

Definition 2.8. A homomorphic aggregate signature is *correct* if for each output of secret-public key pair (sk, pk) of the algorithm $Set(1^\lambda, N)$, the following conditions are valid:

(1) For all $\tau \in \{0,1\}^\lambda$ and $i \in \{1, 2, \ldots, N\}$, if σ is the output of $Sig(sk, \tau, m, i)$, then $Vrf(pk, \tau, m, \sigma) = 1$.

(2) For all $i \in \{1, 2, \ldots, N\}$, if σ_i is the output of $Sig(sk, \tau, m, i)$, then

$$Vrf(\overrightarrow{pk}, \tau, Agg_m(\overrightarrow{pk}, \tau, \overrightarrow{m}, f), Agg_\sigma(\overrightarrow{pk}, \tau, \overrightarrow{\sigma}, f), f) = 1.$$

2.2.2.3 Linearly Homomorphic Aggregate Signatures

The homomorphic aggregate signature schemes present in literature so far ([45] and [74]) are the linearly ones. That is, the computation supported is a linear combination of different messages $m_1, m_2, \ldots, m_N$ coming from different users. Such computation is then reflected on the signatures counterpart. In fact, the final signature σ' joins together the signatures $\sigma_1, \sigma_2, \ldots, \sigma_N$, according to the same linear combinations as the one used for the messages.

In order to derive the linearly homomorphic aggregate signatures from the homomorphic aggregate ones, there are some changes to take into account.

- The messages space $\mathcal{M}$ and the signatures space $\mathcal{Y}$ are the vector space $\mathbb{F}_p^N$ of dimension N defined over the finite field $\mathbb{F}_p$, for a prime number p. Such p is the new output of the algorithm *Set*.
- The messages are vectors $v \in \mathbb{F}_p^N$, i.e. $v = (a_1, a_2, \ldots, a_N)$ where $a_i \in \mathbb{F}_p$.
- If the vectors $v_1, v_2, \ldots, v_N$ are a basis for $\mathbb{F}_p^N$, then the set of admissible functions $f \in \mathcal{F}$ are all the possible linear combinations in the $\mathbb{F}_p$-linear span of $v_1, v_2, \ldots, v_N$.

For a formal definition of linearly homomorphic aggregate signature schemes Definition 2.7 can be adapted. More precisely, the admissible functions are of the form $f = \sum_{i=1}^N c_i v_i$ with respect to the messages and of the form $f = \sum_{i=1}^N c_i \sigma_i$ for the signatures. The same holds for correctness.

Remark 2.3. A formal definition of linearly homomorphic aggregate signature schemes is provided in [74] and [45]. Though, note that in literature the linearly homomorphic aggregate signatures are called homomorphic aggregate signatures. Indeed, the unique examples available so far allow for linear combinations only, that is why "linearly" is omitted. However, we think that it is important to specify whether we are talking about a general homomorphic aggregate signature or a linearly one. Also because in the future, schemes supporting less restrictive functions might be introduced.

Chapter 3
Evaluation of Homomorphic Signature Schemes

Abstract Together with security, there are many other properties that should be taken into account when evaluating a homomorphic signature scheme. In fact it might be important that a signature generated according to an admissible function is indistinguishable from the original ones. In other scenarios a post-quantum signature scheme is needed. In this case we have to make sure that the underlying hardness assumption is expected to face quantum computer attacks. Furthermore, there are situations where computation efficiency and shortness of the generated signatures are important features. In this chapter we discuss and define formally all the above features.

3.1 Hardness Assumptions

In this section the hardness assumptions currently existing homomorphic signature schemes are based on are introduced. First, the hard problems defined on bilinear groups and the Discrete Logarithm Problem are introduced. Afterwards, the hard problems defined on RSA and the Integer Factorization Problem are described. Finally, lattices and the corresponding hardness assumptions are defined. As we will discuss deeply in Sect. 3.1.3, lattice-based assumptions are expected to be resilient against quantum computer attacks.

3.1.1 Bilinear Groups

If $\mathbb{G}$ is an abelian group whose prime order is p and $g \in \mathbb{G}$ is a generator, i.e. $\langle g \rangle = \mathbb{G}$, then the *Discrete Logarithm Problem (DLP)* in $\mathbb{G}$ is the following: given $g, g^a \in \mathbb{G}$, find $a \in \mathbb{Z}_p$. However, several cryptosystems are designed over (security equivalent) weaker variants. The most common are the following.

- *Computational Diffie-Hellman Problem (CDH)*. On the same assumptions as before, given the triple $(g, g^a, g^b) \in \mathbb{G}$, compute g^{ab}.

© The Author(s) 2016

G. Traverso et al., *Homomorphic Signature Schemes*, SpringerBriefs in Computer Science, DOI 10.1007/978-3-319-32115-8_3

- *Decisional Diffie-Hellman Problem (DDH)*. On the same assumptions as before, given the tuple $(g, g^a, g^b, g^c) \in \mathbb{G}$, decide if in $\mathbb{Z}_p$ it is true that $c = ab$, where c is chosen uniformly at random in $\mathbb{Z}_p$.

Many unforgeability proofs for signature schemes rely on hardness assumptions defined over the framework of groups with bilinear maps [7, 18, 52]. Therefore the above problems have to be adapted to this environment. Let us recall briefly the definition.

Definition 3.1. A *bilinear group* is a tuple $(\mathbb{G}_1, \mathbb{G}_2, \mathbb{G}_t, p, e, \varphi)$ such that:

- $\mathbb{G}_1, \mathbb{G}_2$ and $\mathbb{G}_t$ are cyclic groups of prime order p.
- $e : \mathbb{G}_1 \times \mathbb{G}_2 \rightarrow \mathbb{G}_t$ is bilinear, i.e. for all $g_1 \in \mathbb{G}_1, g_2 \in \mathbb{G}_2$ and $a, b \in \mathbb{Z}, e(g_1{}^a, g_2{}^b) = e(g_1, g_2)^{ab}$.
- e is an *admissible* bilinear map, i.e.

  - e is efficiently computable;
  - if g_1 and g_2 are generators of $\mathbb{G}_1$ and $\mathbb{G}_2$, i.e. $\langle g_1 \rangle = \mathbb{G}_1$ and $\langle g_2 \rangle = \mathbb{G}_2$, then $\mathbb{G}_t$ is generated by $e(g_1, g_2)$

- the *Discrete Logarithm Problem* is hard to be computed in $\mathbb{G}_1, \mathbb{G}_2$ and $\mathbb{G}_t$.

The function e is called *bilinear pairing*, or simply *pairing*.

In bilinear pairings two types of hardness assumptions can be distinguished. There are hard problems where $\mathbb{G}_1 \equiv \mathbb{G}_2$ and there are problems where $\mathbb{G}_1$ and $\mathbb{G}_2$ are two distinct groups (see [32]). In the first case, called *symmetric pairing*, both $\mathbb{G}_1$ and $\mathbb{G}_2$ are denoted by $\mathbb{G}$. Here the bilinear map e is of the form

$$e : \mathbb{G} \times \mathbb{G} \rightarrow \mathbb{G}_t.$$

In the second case the map e is of the form

$$e : \mathbb{G}_1 \times \mathbb{G}_2 \rightarrow \mathbb{G}_t$$

and is called *asymmetric pairing*.

Bilinear pairings were employed to attack the Discrete Logarithm Problem (DLP) defined over elliptic curves. For a complete explanation about elliptic curves and their cryptographic usage we refer to [9]. The attack was presented by Menezes et al. in [55] and is therefore called the MOV attack.

Besides this attack, bilinear pairing have been employed in the design of several cryptographic primitives, such as encryption schemes, threshold schemes, and key agreement schemes. With respect to the signature schemes they have been used to design, for instance, blind signatures, multisignatures, and of course homomorphic signature schemes. We refer to [32] for an overview of the possible usages of bilinear pairings.

There are several homomorphic signature schemes whose hardness assumption relies on bilinear groups. Thus, we will now first introduce the Computational

Diffie-Hellman Problem [32]. Afterwards, we present the hardness assumptions based on symmetric bilinear maps followed by the hardness assumptions based on asymmetric bilinear maps.

Definition 3.2. Given a cyclic group $\mathbb{G}$ of prime order p, the *Computational Diffie-Hellman Problem (CDH)* in $\mathbb{G}$ is the following: given the group elements $g, g^a, g^b \in \mathbb{G}$, compute $g^{ab} \in \mathbb{G}$.

However, in bilinear groups the decisional Diffie-Hellman problem is easy to solve [49] and therefore it cannot be used. For such situations, Boneh et al. introduced in [16] the Decisional Linear Problem.

Definition 3.3. Let $\mathbb{G}$ be a cyclic group of prime order p and g a generator of $\mathbb{G}$, i.e. $\langle g \rangle = \mathbb{G}$. Furthermore, let e be a symmetric bilinear map $e : \mathbb{G} \times \mathbb{G} \to \mathbb{G}_t$. Given other arbitrary u, v, h, all of them generators of $\mathbb{G}$, i.e. $\langle u \rangle = \langle v \rangle = \langle h \rangle = \mathbb{G}$. Then the *Decision Linear Problem (DLIN)* in $\mathbb{G}$ is the following: having $a, b, c \in \mathbb{Z}_p^*$ and given $u, v, h, u^a, v^b, h^c \in \mathbb{G}$ as input, output *yes* if $a + b = c$ and *no* otherwise.

Another problem is the Flexible Diffie-Hellman Problem, which is slightly stronger than the standard Diffie-Hellman Problem. However it is still a *simple* one, since it implies the fact that distinguishing g^{abc} from a randomly-given tuple (g, g^a, g^b, g^c) is hard [7].

Definition 3.4. Assume we have a cyclic group $\mathbb{G}$ of order p, generator g, i.e. $\langle g \rangle = \mathbb{G}$ and a symmetric bilinear map $e : \mathbb{G} \times \mathbb{G} \to \mathbb{G}_t$. Then the *Flexible Diffie-Hellman Problem (flexDHP)* is the following: given the triple (g, g^a, g^b), where $a, b \in \mathbb{Z}_p$, find another triple $(g^\mu, g^{a \cdot \mu}, g^{ab \cdot \mu}) \in \mathbb{G}^3$, where $\mu \neq 0$.

A variant of the above problem is called q-Simultaneous Flexible Pairing Problem [1].

Definition 3.5. For a group $\mathbb{G}$ of order p and generators g_z, h_z, g_r, h_r, i.e. $\langle g_z \rangle = \langle h_z \rangle = \langle g_r \rangle = \langle h_r \rangle = \mathbb{G}$, and symmetric bilinear map $e : \mathbb{G} \times \mathbb{G} \to \mathbb{G}_t$ the *q-Simultaneous Flexible Pairing (q-SFP)* is the following: given a tuple $(g_z, h_z, g_r, h_r, a, \overline{a}, b, \overline{b}) \in \mathbb{G}^8$ and a set of q tuples $(z_j, r_j, s_j, t_j, u_j, v_j, w_j) \in \mathbb{G}^7$ such that

(1) $e(a, \overline{a}) = e(g_z, z_j) \cdot e(g_r, r_j) \cdot e(s_j, t_j)$,
(2) $e(b, \overline{b}) = e(h_z, z_j) \cdot e(h_r, u_j) \cdot e(v_j, w_j)$,

the goal is to find another fresh tuple $(z', r', s', t', u', v', w') \in \mathbb{G}^7$, where $z' \notin \{1_\mathbb{G}, z_1, \ldots, z_q\}$, for which (1) and (2) are still valid.

If in the bilinear group $(\mathbb{G}_1, \mathbb{G}_2, \mathbb{G}_t, p, e, \varphi)$ the groups $\mathbb{G}_1$ and $\mathbb{G}_2$ are distinct, the bilinear map e becomes $e : \mathbb{G}_1 \times \mathbb{G}_2 \to \mathbb{G}_t$. In the following the two hardness assumptions defined over this scenario are provided, starting by the co-Computational Diffie-Hellman Problem [18].

Definition 3.6. Let $\mathbb{G}_1, \mathbb{G}_2$ be groups of order p, let g_1 be the generator of $\mathbb{G}_1$ and g_2 the generator of $\mathbb{G}_2$, i.e. $\langle g_1 \rangle = \mathbb{G}_1$ and $\langle g_2 \rangle = \mathbb{G}_2$, and $e : \mathbb{G}_1 \times \mathbb{G}_2 \to \mathbb{G}_t$

be an asymmetric bilinear map. The *co-Computational Diffie-Hellman (co-CDH)* is the following: given $g_1, g_1{}^a \in \mathbb{G}_1$ and $g_2, g_2{}^b \in \mathbb{G}_2$, where $a, b \in \mathbb{Z}_p$, compute $g_2{}^{ab}$.

We now define the q-Strong Diffie-Hellman Problem [12], which was introduced by Boneh and Boyen.

Definition 3.7. Let $(\mathbb{G}_1, \mathbb{G}_2)$ be groups of order p, with generators g_1, g_2, i.e. $\langle g_1 \rangle = \mathbb{G}_1$ and $\langle g_2 \rangle = \mathbb{G}_2$, and $e : \mathbb{G}_1 \times \mathbb{G}_2 \to \mathbb{G}_t$ be an asymmetric bilinear map. Then the *q-Strong Diffie-Hellman Problem (q-SDH)* is the following: given the tuple $(g_1, g_1{}^x, g_1{}^{x^2}, \ldots, g_1{}^{x^q}, g_2, g_2{}^x)$ for $x \in \mathbb{Z}_p \backslash \{0\}$, compute a pair $(c, g_1^{\frac{1}{x+c}})$, where $c \in \mathbb{Z}_p \backslash \{0\}$.

3.1.1.1   Other Assumptions

There are five further complexity assumptions proposed in [5] and [6]. They are defined over the following framework, which is slightly different from the one used so far.

Assume the case that $\mathbb{G}_1 = \mathbb{G}_2 = \mathbb{G}$, therefore $e : \mathbb{G} \times \mathbb{G} \to \mathbb{G}_t$ is a symmetric bilinear map. The order p of $\mathbb{G}$ is given by $p = p_1 p_2 p_3$, where p_1, p_2, p_3 are prime numbers. Let $\mathbb{G}_{p_i}$ be the subgroup of order p_i, for $i \in \{1, 2, 3\}$ and let $\mathbb{G}_{p_i p_j}$ be the subgroup of order $p_i p_j$, for $i \neq j$.

Remark 3.1. Let $(u, v) \in \mathbb{G}_{p_i p_j}$ of order p_i and p_j, respectively. Then $e(u, v) = 1_{\mathbb{G}_t}$ [5].

- **Ass. 1** For given elements $g \in \mathbb{G}_{p_1}, X_3 \in \mathbb{G}_{p_3}$ and a group element T, it is hard to decide whether $T \in \mathbb{G}_{p_1 p_2}$ or $T \in \mathbb{G}_{p_1}$.
- **Ass. 2** Let us assume that $g, X_1 \in \mathbb{G}_{p_1}, X_2, Y_2 \in \mathbb{G}_{p_2}$, and $Y_3, Z_3 \in \mathbb{G}_{p_3}$. Then it is hard to decide, given a tuple $(g, X_1 X_2 Z_3, Y_2 Y_3)$ and T, if $T \in (G)$ or $T \in \mathbb{G}_{p_1 p_3}$.
- **Ass. 3** Let us assume that $g \in \mathbb{G}_{p_1}$, $X_2, Y_2, Z_2 \in \mathbb{G}_{p_2}$, $X_3 \in \mathbb{G}_{p_3}$ and, $\alpha, s \in \mathbb{Z}_p$. Then given the tuple $(g, g^\alpha X_2, X_3, g^s Y_2, Z_2)$ it is hard to compute $e(g, g)^{\alpha s}$.
- **Ass. 4** Let us assume that, for $t \in \mathbb{Z}_p$, the elements $g, w, g^t, X_1 \in \mathbb{G}_{p_1}, X_2, Y_2, Z_2 \in \mathbb{G}_{p_2}$, and $X_3, Y_3, Z_3 \in \mathbb{G}_{p_3}$ are given. Having the element $T \in \mathbb{G}$ and the tuple $(g, w, g^t, X_1 X_2, X_3, Y_2 Y_3)$ it is hard to decide if $T = w^t Z_3$ or $T = w^t Z_2 Z_3$.
- **Ass. 5** Let us assume that $a, b, c \in \mathbb{Z}_p$, $g \in \mathbb{G}_{p_1}$, $X_2, Y_2, Z_2 \in \mathbb{G}_{p_2}$, and $X_3 \in \mathbb{G}_{p_3}$. Then given the tuple $(g, g^a, g^b, g^{ab} X_2, X_3, g^c Y_2, Z_2)$ it is hard to compute $e(g, g)^{abc}$.

3.1.2   RSA

A well known and widely used cryptosystem is RSA, proposed by Rivest et al. [60]. In general, the RSA assumption implies that the *Integer Factorization Problem* is hard, while the converse still has to be proven. There are two hardness assumptions that are defined within the RSA scenario: the Standard RSA problem [73] and the Strong RSA problem [22]. Before describing them, some useful definitions used within the RSA framework are recalled.

Definition 3.8. An integer N is called an *RSA modulus* if it holds that $N = pq$, where p, q are two distinct odd prime numbers.

Definition 3.9. The *Euler's phi function* of an integer $n > 0$ is denoted by $\varphi(n)$ and defined as

$$\varphi(n) := \#\{k \in \mathbb{Z} \quad \text{such that} \quad \gcd(n, k) = 1, \quad 1 \leq k \leq n\},$$

where $\gcd(n, k)$ is the greatest common divisor of the integers and $n, k \in \mathbb{Z}$.

Euler's phi function is multiplicative, i.e. $\varphi(nm) = \varphi(n)\varphi(m)$, where m, n are positive integers. If the input to the Euler's phi function is a prime number p, it holds that $\varphi(p) = p - 1$. Using the multiplicative property, it holds that

$$\varphi(pq) = \varphi(p)\varphi(q) = (p - 1)(q - 1).$$

The Standard RSA problem is defined as follows.

Definition 3.10. Let (N, e) be a pair of integers, where $N = pq$ for some odd prime numbers $p, q, e \in \mathbb{Z}_N \setminus \{1\}$ and $\gcd(e, \varphi(N)) = 1$. Given an element $z \in \mathbb{Z}_N$, the *Standard RSA* problem is to compute the integer y such that $y^e \equiv z \mod N$.

The Strong RSA problem is defined as follows [48].

Definition 3.11. Let (N, e) be a pair of integers, where $N = pq$ for some odd prime numbers $p, q, e \in \mathbb{Z}_N \setminus \{1\}$ and $\gcd(e, \varphi(N)) = 1$. Given an element $z \in \mathbb{Z}_N$ and the freedom to choose e, the *Strong RSA* problem is to compute the integer y such that $y^e \equiv z \mod N$.

Remark 3.2. The difference between the Standard RSA problem and the Strong RSA problem is that in the first case the exponent e is chosen independently of z, while in the second case, given the freedom of choice to the attacker, e might be chosen dependent on z. See [29] and [48] for further information.

3.1.3 Lattices

Peter Shor proved that the Discrete Logarithm Problem and the Integer Factorization Problem can be solved in polynomial time if the attack is run on a quantum computer [63]. Cryptographic primitives relying on such problems may then be broken once quantum computers will be available. Lattices are an important mathematical construction as the hard problems defined on them are expected to face quantum computer attacks [8]. Thus, lattice-based cryptographic primitives are assumed to provide security even in the long-term perspective. In the following, the definition of lattices is provided together with a description of their most important properties. For a more complete overview of lattices we refer to [8].

Having a positive integer $n > 0$ and $\mathbb{R}^n$ as underlying vector space lattices are defined as follows [8].

Definition 3.12. Let $b_1, b_2, \ldots, b_n \in \mathbb{R}^n$ be a set of linearly independent vectors. The *lattice* Λ (generated by $b_1, b_2, \ldots, b_n$) is defined as:

$$\Lambda := \mathscr{L}(b_1, b_2, \ldots, b_n) = \{\sum_{i=1}^{n} x_i b_i \quad \text{such that} \quad x_1, x_2, \ldots, x_n \in \mathbb{Z}\}.$$

In other words, the lattice Λ is the set of all linear combinations of $b_1, b_2, \ldots, b_n$ with coefficients in $\mathbb{Z}$.

As the lattice is defined by elements of a vector space (in this case $\mathbb{R}^n$), a definition of "a basis" for Λ and "the dimension" of Λ can be provided.

- A *basis* for the lattice Λ is any set of linearly independent vectors that generates Λ itself. Referring to the definition above, $\mathscr{B} := \{b_1, b_2, \ldots, b_n\}$ can be denoted as a basis for the lattice Λ.
- The *dimension* of the lattice Λ is the number of linearly independent vectors that forms a basis for the lattice.

In addition, a lattice Λ can be defined by a matrix B, whose columns are the vectors $b_1, b_2, \ldots, b_n$.

$$\Lambda = \mathscr{L}(B) = \{Bx : x \in \mathbb{Z}^n\}.$$

In this case the *determinant* of the lattice Λ is the absolute value of the determinant of the matrix B, i.e. of the matrix representing one of its basis:

$$\det(\Lambda) = |\det(B)|.$$

In cryptography, a particular family of lattices is usually employed, called the *q-ary lattices*, where q is a positive prime integer. This means that if the lattice Λ is generated by $b_1, b_2, \ldots, b_n$, then an element $x = x_1 b_1, x_2 b_2, \ldots, x_n b_n \in \Lambda$ is determined by $x \mod q$. For the cryptographic usage we are interested in q-ary lattices where q is much smaller than $|\det(B)|$ (see [8]).

Let us consider the integers $n, m,$ and q and a matrix $A \in \mathbb{Z}_q^{n \times m}$. Then, two q-ary lattices of dimension m can be defined as follows.

- $\Lambda_q(A) := \{y \in \mathbb{Z}^m \quad \text{such that} \quad y = A^T s \mod q, s \in \mathbb{Z}^n\}$.
- $\Lambda_q^{\perp}(A) := \{y \in \mathbb{Z}^m \quad \text{such that} \quad Ay = 0 \mod q\}$.

Then, for any vector $x \in \mathbb{R}^n$ such that $Ay = x \mod q$ we define the *shifted lattice* as

$$\Lambda_x^\perp := \{y \in \mathbb{Z}^m \quad \text{such that} \quad Ay = x \mod q\}.$$

In the following the most common hard problems defined in the lattice framework are discussed. These are the Shortest Vector Problem and the Closest Vector Problem.

- The Shortest Vector Problem (SVP) describes finding a vector $x \neq 0$ such that it is the shortest vector in the lattice Λ with a basis $B = \{b_1, b_2, \ldots, b_n\}$.
- The Closest Vector Problem (CVP) denotes finding a vector x in the lattice Λ such that it is the closest with respect to another given vector (non necessarily belonging to the lattice itself).

In lattice-based cryptography, many primitives are actually built on approximation variants of the SVP and CVP, where the approximation factor is denoted by γ.

Finally the hardness assumptions lattice-based homomorphic signature schemes are built on are defined. Assume an arbitrary matrix $A \in \mathbb{Z}_q^{n \times m}$ is given, then the following hardness assumption can be defined on the lattice $\Lambda_q^\perp(A)$ [45].

Definition 3.13. Given a uniform and random matrix $A \in \mathbb{Z}_q^{n \times m}$ for positive integers m, n, q and given a real number $\beta \in \mathbb{R}$, the *Small Integer Solution (SIS)* is the following problem: find a nonzero integer vector $y \in \Lambda_q^\perp(A)$ such that $Ay = 0$ mod q and $||y|| \leq \beta$.

A variant of the above hardness assumption is the k-Small Integer Solution Problem.

Definition 3.14. Given a matrix $A \in \mathbb{Z}_q^{n \times m}$, a real number $\beta \in \mathbb{R}$, and k short vectors $y_1, y_2, \ldots, y_k \in \mathbb{Z}^m$ such that $A \cdot y_i = 0 \mod q$ for any $i \in \{1, 2, \ldots, k\}$, the *$k$-Small Integer Solution (k-SIS)* problem consists of finding another short vector $y \in \mathbb{Z}^m \backslash \mathbb{Q} - span\{x_1, x_2, \ldots, x_k\}$ such that $A \cdot y = 0 \mod q$ and $||y|| \leq \beta$.

Again, another variant of the SIS problem is defined, that is the Inhomogeneous Small Integer Solution Problem. It was introduced in [38] and consists of finding a short solution to a random inhomogeneous system.

Definition 3.15. Given a uniformly random integer q, a uniformly random matrix $A \in \mathbb{Z}_q^{n \times m}$, a syndrome $x \in \mathbb{Z}_q^n$, and a real number $\beta \in \mathbb{R}$, the *Inhomogeneous Small Integer Solution (ISIS)* is the following problem: find a vector of integers $y \in \mathbb{Z}^m$ such that $Ay = x \mod q$ and $||y|| \leq \beta$.

3.2 Efficiency and Size

Efficiency is a property which still has to be clearly defined in the framework of the homomorphic signature schemes. Indeed, there is no precise standard with respect to comparing all the existing schemes in a rigorous and unique way. Though, it would be desirable as a future work. However, some information and partial comparisons are available in the state of the art, as we will discuss in Chap. 4.

Moreover, the notion of "*succinctness*" for a signature's size, i.e. the desirable length for a signature, is commonly accepted. For a fixed security parameter λ, a (homomorphic) signature scheme is called *succinct*, if the signature's length depends only logarithmically on the size N of the data set [14].

3.3 Security

While all the digital signature schemes are considered secure when they are unforgeable under adaptive-chosen message attacks (see Sect. 1.2.3), in the homomorphic signature schemes framework, there are two possible meanings for "secure". Basically a homomorphic signature scheme is unforgeable if an adversary is not able to generate a valid message-signature pair, when the message has not been already signed and the messages cannot be derived from data previously seen. In this scenario we distinguish between the *weak adversary* (see Sect. 3.3.1) and the *strong adversary* (see Sect. 3.3.2). In the description of the currently existing homomorphic signature schemes in Chap. 4, it will be highlighted in which scenario the scheme in question is secure.

3.3.1 Weak Adversary

In the weak adversary scenario the attacker A cannot freely choose the messages signed by a signing oracle $\mathcal{O}_S$. That is, A cannot choose *any* message at *any* time. It is instead restricted to make one query on a sequence of data sets of its choice. $\mathcal{O}_S$ randomly chooses a tag τ for each data set defining to which messages a signature will be provided. This notion of security is formalized by Boneh et al. in [18]. However, the game described in [18] is for linearly homomorphic signature schemes. Thus, we refer to the more general description given by Boneh and Freeman in [14].

Let us recall that a homomorphic signature scheme is the tuple (*Set*, *Sig*, *Vrf*, *Eval*), λ is the security parameter, and N the maximum data set size.

Definition 3.16. A homomorphic signature scheme (*Set*, *Sig*, *Vrf*, *Eval*) with security parameter λ is *unforgeable* if for all N and for all polynomials $t(\cdot)$ the probability for a probabilistic, polynomial-time adversary A to succeed in the following game is negligible with respect to λ.

1) A key pair (sk, pk) is produced by the algorithm $Set(1^\lambda, N)$ and a set $\mathcal{F}$ of admissible functions $f : \mathcal{M}^N \to \mathcal{M}$ is defined.
2) The attacker A knows pk and can request signatures from the signing oracle $\mathcal{O}_S$ in the following way. A selects a sequence of data sets $\vec{m}_i \in \mathcal{M}^N$. For each

$i \in \{1, 2, \ldots, N\}$ $\mathscr{O}_S$ chooses a tag $\tau_i \in \{0, 1\}^\lambda$, uniformly at random. Then, $\mathscr{O}_S$ returns to A τ_i and the signatures σ_{ij}, where $j \in \{1, 2, \ldots, N\}$ and σ_{ij} is the output of $Sig(sk, \tau_i, m_{ij}, j)$.

3) The attacker A outputs a tuple $(\tau^*, m^*, \sigma^*, f)$.
4) A succeeds if $Vrf(pk, \tau^*, m^*, \sigma^*, f^*) = 1$ and either

- $\tau^* \neq \tau_i$ for all $i \in \{1, 2, \ldots, N\}$ (*Type I Forgery*), or
- $\tau^* = \tau_i$ for a certain i and $m^* \neq f^*(\vec{m}_i)$ (*Type II Forgery*).

The above game formalizes the attacker's intent. Instead of aiming at getting a new pair (m^*, σ^*) as for the digital signatures' case, this time the attacker wants to output a triple (m^*, σ^*, f) (where σ^* is the signature over $f(m^*)$) which cannot be derived from data and signatures previously seen.

Remark 3.3. Without the tag τ, the definition of security would be weaker. Indeed without such a file identifier, the adversary would just be able to query some messages on $\mathscr{M}$ and not on an entire data set (see [73]).

3.3.2 Strong Adversary

In [34], Freeman strengthens the adversary control over the messages that are signed by the signing oracle $\mathscr{O}_S$. A is not restricted to query all the messages belonging to a given data set at one time. Now the attacker can query *one* message at a time, and choose the following one based on the output of the previous query. Furthermore, it can do this adaptively within each data set and spread the queries among the data sets. In this way the attacker can win in a third way: "Type III forgery". That is, A might output a triple (m^*, σ^*, f) where σ^* is the signature to the pair (m^*, f) which corresponds to a previously seen data set. Though, the adversary has not queried enough messages on that data set in order to simulate the behavior of f. It follows a more detailed description of the game provided in [34].

Definition 3.17. A homomorphic signature scheme (*Set, Sig, Vrf, Eval*) with security parameter λ is *unforgeable* if for all N and for all polynomials $t(\cdot)$, the probability for a probabilistic, polynomial-time adversary A to succeed in the following game is negligible with respect to λ.

1) A key pair (sk, pk) is produced by the algorithm $Set(1^\lambda, N)$ and a set $\mathscr{F}$ of the admissible functions $f : \mathscr{M}^N \to \mathscr{M}$ is defined.
2) The attacker A can request signatures from the signing oracle $\mathscr{O}_S$ in the following way. The adversary A selects a filename $F \in \{0, 1\}^*$ and a message $m \in \mathscr{M}$. If the message m is the first request with respect to F, $\mathscr{O}_S$ selects a tag $\tau_F \in \{0, 1\}^\lambda$ uniformly at random. Then, $\mathscr{O}_S$ sets a counter $i_F = 1$ and returns τ_F to A. If the message m is not the first request with respect to F, then $\mathscr{O}_S$ checks the value of τ_F and increments the counter i_F by 1.

3) $\mathcal{O}_S$ computes the signature $\sigma^{(F,i_F)}$, output of the signing algorithm $Sig(sk, \tau_F, m, i_F)$. The signature $\sigma^{(F,i_F)}$ is given to A.
4) Steps 3)–4) are repeated for a polynomial number of times. The only constraint is that at most N messages can be queried for each filename F.
5) The adversary A outputs a tuple $(\tau^*, m^*, \sigma^*, f)$.

Before concluding with Step 6) and stating how the attacker A wins, the following definition is introduced [34].

Definition 3.18. In the framework introduced in Definition 3.17, a function $f : \mathcal{M}^N \to \mathcal{M}$ is *well defined* on the filename $F \in \{0, 1\}^*$ if either:

- $i_F = N$, or
- $i_f < N$ and $f(M_F, m_{i_F+1}, m_{i_F+2}, \ldots, m_N)$ takes the same value for all possible choices of $(m_{i_F+1}, m_{i_F+2}, \ldots, m_N) \in \mathcal{M}^{N-i_F}$, where M_F is the tuple of messages m queried for the filename F, listed in the order they were requested.

Now we can conclude the definition of unforgeability against the strong adversary.

6) The adversary A succeeds if $Vrf(pk, \tau^*, m^*, \sigma^*, f^*)$ and either

  - $\tau^* \neq \tau_F$ for all filenames F queried by A (*Type I Forgery*), or
  - $\tau^* = \tau_F$ for filename F, f^* is well defined on F, and $m^* \neq f^*(M_F)$ (*Type II Forgery*), or
  - $\tau^* = \tau_F$ for filename F and f^* is not well defined on F (*Type III Forgery*).

3.4 Privacy

In many practical applications it is necessary to provide a certain level of privacy. There are three different notions of privacy, according to the level of protection achieved by a scheme.

Let us call $\sigma_1, \sigma_2, \ldots, \sigma_N$ the set of signatures from which a signature σ' for a message m' is derived. A homomorphic signature scheme is said to be *weakly context hiding* if σ' only reveals information about the corresponding message m', but does not leak any information about the data set $m_1, m_2, \ldots, m_N$ of the respective input signatures.

This notion of privacy has been introduced in [4], together with its stronger version: *strong context hiding*. This privacy level is achieved by a signature scheme when it is not possible to distinguish whether the signature σ' has been computed out of the signatures $\sigma_1, \sigma_2, \ldots, \sigma_N$ or whether it has been generated directly to a message m'. This privacy level requires the infeasibility of linking the signature σ' to the original ones $\sigma_1, \sigma_2, \ldots, \sigma_q$, even if they are public knowledge (see [45]).

According to the authors of [6], the definition in [4] takes only the indistinguishability from honestly generated signatures into account. It does not imply unlinkability when the original signatures are chosen by an attacker. In order to address this, in [6] a new notion of privacy, *completely context hiding*, is defined. In fact this notion even requires (statistical) context hiding on signatures in the case that the input signatures of the function are chosen by an adversary having access to the secret key (see [45]).

To conclude, there are three notions of privacy for homomorphic signature schemes. They are listed in hierarchical order and the previous one implies the following ones:

- completely context hiding;
- strong context hiding;
- weakly context hiding.

3.5 Random Oracle Model vs. Standard Model

Hash functions are an important primitive for the design of signature schemes, both the digital and the homomorphic ones. The assumptions made with respect to hash functions employed determine whether the signature scheme in question is secure in the *Random Oracle Model* or in the *Standard Model*. Before describing these two frameworks, we briefly explain what a hash function is. For a more formal definition we refer to [48].

A *hash function H* is a function that compresses its input and returns an output of fixed length. Furthermore, a hash function satisfies the following properties:

- it is hard to find two inputs x_1, x_2 such that $H(x_1) = H(x_2)$ and $x_1 \neq x_2$ (*collision resistance*);
- given an input x_1, it is hard to find an input x_2 such that $H(x_1) = H(x_2)$ and $x_1 \neq x_2$ (*second pre-image resistance*).

The main distinction between the Random Oracle Model and the Standard Model is how the security proof is performed. Some schemes need to be proven secure in an ideal framework. This means that the hash function employed is assumed to behave like a perfectly random function, called *random oracle*. In this case the scheme is said to be secure in the Random Oracle Model. There are other schemes that can be set instead in a more realistic scenario, called Standard Model. In this case perfectly randomness is not necessary to prove the security of the signature scheme and that is considered a valuable property.

Chapter 4
State of the Art of Homomorphic Signature Schemes

Abstract In this chapter the state of the art with respect to homomorphic signature schemes is presented. Due to the large number and the different properties they satisfy, they are discussed in separate groups, according to the computations they support. The linearly homomorphic signature schemes are further divided with respect to the hardness assumption they rely on. Afterwards, the existing homomorphic signature schemes for polynomial functions and the fully homomorphic ones are described. Regarding the existing homomorphic signature schemes for the multi-users case, the linearly homomorphic aggregate signature schemes and the multiple sources linearly homomorphic signature schemes are presented separately. The investigated properties are the ones introduced in the previous section. For each scheme the underlying hardness assumption is specified, then we provide information about the efficiency of the schemes and their signature's length. Afterwards, the general safety of the scheme is discussed: which adversary the signature can cope with and which level of privacy it achieves.

4.1 Linearly Homomorphic Signature Schemes Defined Over Bilinear Groups

The first homomorphic signature schemes introduced were the linearly ones. The earliest signature we present is the one proposed by Boneh et al. in 2009 in [18], but it is not the first one in literature. Actually, there are other schemes relying on Diffie-Hellman problems and bilinear groups that had been published before (see [24] and [76]). However, they have already been proven to be not practical or even not completely secure [31, 64, 73]. That is why we do not discuss them in this work. Furthermore, we do not include the recent work by Libert et al. [53], because their scheme is not linear in the sense of vector fields. The construction of linearly homomorphic signature schemes over bilinear groups exploits the homomorphic property of DLP, from which bilinear groups are defined (see Sect. 3.1.1). In fact, DLP is based on operations on exponentiations, i.e. $f^a \cdot g^a = (f \cdot g)^a$ for $a, f, g \in \mathbb{Z}$. It follows that the sum of two bilinear maps and the product of a scalar and a bilinear map is again bilinear. Thus, bilinear maps form a vector space and allow for linear operations. In the following, existing solutions and further properties are provided.

© The Author(s) 2016

G. Traverso et al., *Homomorphic Signature Schemes*, SpringerBriefs in Computer Science, DOI 10.1007/978-3-319-32115-8_4

4.1.1 Signing a Linear Subspace: Signature Schemes for Network Coding, by Boneh et al. [18]

Boneh et al.'s work [18] is the milestone of linearly homomorphic signatures. Indeed it is considered to be the first one to provide a practical framework for such schemes and it introduces the notion of weak adversaries. The scheme proposed is proven secure assuming that the co-CDH problem in $(\mathbb{G}_1, \mathbb{G}_2)$ is hard to solve. The bilinear pairing assigned to this hardness assumption is asymmetric. Referring to the notations in Sect. 3.1.1 the public key contains the description of the bilinear group itself and two elements of the group $\mathbb{G}_2$. The signature consists of one single element of the group $\mathbb{G}_1$, while the verification algorithms requires two pairing operations. The scheme is claimed to have low communication overhead, because of the independence of both public key and signature's sizes with respect to the maximum number of messages within the data set (see [73]). Furthermore, a lower bound on signatures' length is provided. The security is proven in the Random Oracle Model only. However, this scheme can cope with the weak adversary and the signature enjoys the property to be completely context hiding in terms of privacy, as it has been later pointed out in [4].

4.1.2 Homomorphic Network Coding Signatures in the Standard Model, by Attrapadung and Libert [5]

The signature scheme described in [5] is the first linearly homomorphic signature scheme proven in the Standard Model. The scheme is defined over bilinear groups, where the decisional assumptions Ass. 1 and Ass. 2 and the computational assumption Ass. 3 are employed to provide unforgeability. The bilinear pairing assigned to these hardness assumptions is symmetric. In this case, instead of working with prime fields (like for the scheme described in [18]), the coordinates have to be chosen in $\mathbb{Z}_n$, where n is the composite order of the bilinear groups. Referring to the notation used in Sect. 3.1.1, the public key is composed of the description of the bilinear group, $N + 3$ elements $g, u, v, h_1, \ldots, h_N$ of the group $\mathbb{G}_1$, the value $e(g, g)^\alpha$ for a random $\alpha \in \mathbb{Z}_n$, and a group element of $\mathbb{G}_3$. The signature consists of three group elements of $\mathbb{G}_1$ and four pairings are needed during the verification. Furthermore, like in [18], the unforgeability of the scheme can face the weak adversary, but it is proven secure in the Standard Model. Note that, according to [34], it is possible to modify the proof in the Standard Model and to face even the strong adversary. Another valuable property is that the scheme is strongly context hiding.

4.1.3 Computing on Authenticated Data: New Privacy Definitions and Constructions, by Attrapadung et al. [6]

The same authors of [5] proposed 1 year later a similar scheme presented in [6]. The groups $(\mathbb{G}, \mathbb{G}_t)$ considered are of composite order $p = p_1 p_2 p_3$, the hardness assumptions are Ass. 1, Ass. 2 , Ass. 4 and Ass. 5 and the bilinear pairings used are symmetric. If we compare this scheme with the one presented in [5], then the public key is the same except that in [6] the term $e(g, g)^\alpha$ is replaced by g^α. The signature is composed of two group elements of $\mathbb{G}_1$, that is 33 % shorter then the signature in [5]. This reduces the number of pairings performed during the verification to two. This scheme is proven secure against the weak adversary in the Standard Model and achieves strongly context hiding privacy.

4.1.4 Efficient Network Coding Signatures in the Standard Model, by Catalano et al. [22]

The scheme presented in [22] is proven secure under the q-SDH assumption and uses asymmetric bilinear pairings. Referring to the notations in Sect. 3.1.1, the signature is composed of one group element of $\mathbb{G}_1$ and one element in $\mathbb{Z}_p$. The verification algorithm involves the computation of two pairings. The scheme is proven secure under the weak adversary in the Standard Model. However, privacy has not be clarified yet.

4.1.5 Improved Security for Linearly Homomorphic Signatures: A Generic Framework, by Freeman [34]

In [34], Freeman provides a generic framework that allows to transform non-homomorphic signatures into their respective homomorphic counterparts. Specifically, the author used the proposed algorithm to convert several of them and prove their security in the Standard Model, maintaining the original hardness assumptions they were relying on. That is, the scheme proposed in [67] under the CDH and the scheme proposed in [11] under the q-Strong DH. Note that the scheme relying on the CDH assumption uses symmetric bilinear pairings, while the scheme relying on the q-Strong DH uses asymmetric pairings. The public key is composed of the public key for the original algorithm and some additional elements. Using the notion given in Sect. 3.1.1 the public key consists also of $2N + 1$ group elements $h_1, h_2, \ldots, h_N, t_1, t_2, \ldots, t_N, u \in \mathbb{G}_1$, the ring of the exponents of $\mathbb{G}_1$, and two bounds on $\mathbb{G}_1$'s size, denoted by $|\mathbb{G}_1|$. The signature is composed of the signature (σ_1, σ_2) of the respective non-homomorphic scheme, a group element σ_3 of $\mathbb{G}_2$, and an integer $s \in \mathbb{Z}_{|\mathbb{G}_1|}$. The verification is performed in three steps. First of all, it is verified whether the signature (σ_1, σ_2) is actually a valid signature for the non-homomorphic scheme. Then it is checked whether the other two components σ_3 and s are consistent with the signature (σ_1, σ_2). Finally, it is checked if the bounds on $|\mathbb{G}_1|$ are satisfied or not. Note that the first step in the verification algorithms depends on how the verification of the respective non-homomorphic scheme is performed. In the non-homomorphic signature scheme presented in [67], there are three pairings in the verification algorithm, while in the non-homomorphic signature scheme described in [11] there are two pairings. However, in both cases one of the pairings is $e(g_1, g_2)$, which is computed at the initialization time once for all and then cached. We can then assume that the number of pairings are one for the scheme presented in [67] and two for the scheme introduced in [11].

Both schemes achieve a high security level: the unforgeability is proven secure against the stronger adversary in the Standard Model. Furthermore, the signatures are weakly context hiding.

4.1.6 Efficient Completely Context-Hiding Quotable and Linearly Homomorphic Signatures, by Attrapadung et al. [7]

The work presented in [7] by Attrapadung et al. is a linearly homomorphic signature scheme relying on the Flex-DH and the DLIN hardness assumptions (defined in Sect. 3.1.1) and employing symmetric pairings. These hardness assumption allow to improve the signature scheme described in [6] (which is weakly context hiding) in order to make it completely context hiding. Referring to the notation used in Sect. 3.1.1, the public key consists of the description of the bilinear group, the group elements $g, u, g^\alpha, g_1, g_2, \ldots, g_N, u_o, u_1, \ldots, u_N \in \mathbb{G}$, where $\alpha \in \mathbb{Z}_p$ is random. The signature consists of six elements of the group $\mathbb{G}^{16}$ and the verification needs the computation of twenty-four pairings. The scheme is proven secure in Standard Model and can cope with the weak adversary. As mentioned before this scheme is completely context hiding.

4.1.7 Secure Network Coding Against Intra/Inter-Generation Pollution Attacks, by Guangjun and Bin [42]

The homomorphic signature scheme proposed by Guangjun and Bin [42] is based on the homomorphic signature scheme presented in [18]. However, the hardness assumption employed is the CDH rather then the co-CDH, which is the problem used in [18]. The signature consists of one group element. Despite the fact that the hardness assumption involves bilinear groups, no pairing is involved in the verification algorithm which requires only exponentiations. In this way, the

Table 4.1 Linearly homomorphic signature schemes defined over bilinear groups

Signatures	Hard. Ass	Adversary	Privacy	Model	Efficiency	Pairing
Section 4.1.1	co-CDH	Weak	Complete	ROM	2 Pairings	Asymm.
Section 4.1.2	Ass. 1,2,3	Weak	Strong	Standard	4 Pairings	Symm.
Section 4.1.3	Ass. 1,2,4,5	Weak	Strong	Standard	2 Pairings	Symm.
Section 4.1.4	q-SDH	Weak	$\emptyset$	Standard	2 Pairings	Asymm.
Section 4.1.5	CDH/q-Strong DH	Strong	Weak	Standard	1/2 Pairings	Symm./asymm.
Section 4.1.6	Flex DH/DLIN	Weak	Complete	Standard	24 Pairings	Symm./symm.
Section 4.1.7	CDH	Weak	$\emptyset$	ROM	0 Pairings	–

verification is supposed to be fast as the most expensive operation is missing. The security is proved in the Random Oracle Model and the scheme can cope with the weak adversary. Privacy has not been discussed yet. However, it has been shown in [26] that a forgery can be output by an attacker with high probability after several signatures have been generated. That is due to the fact that the adversary can obtain linear combinations of the secret keys. Thus, the authors of [26] show how to overcome this drawback and make the homomorphic signature scheme immune to such attack.

4.1.8 Summary of Linearly Homomorphic Signature Schemes Defined over Bilinear Groups

In Table 4.1 we summarize the properties of the linearly homomorphic signatures discussed so far. Regarding efficiency, we highlight the number of pairings during the verification process, as they are the most expensive computations involved. We see that only one scheme, the one presented in Sect. 4.1.5, is secure against the strong adversary. However, with 1/2 pairings this scheme is quite efficient, it provides weak privacy, and its security has been proven in the Standard Model. Signature schemes that provide strong or complete privacy can cope with the weak adversary only.

4.2 RSA-Based Linearly Homomorphic Signature Schemes

The earliest signature relying on RSA we present is the one proposed by Gennaro et al. in 2010 [36], but it is not the first one in literature. Actually, there are other schemes relying on RSA that had been published before (see [71] and [72]). However, they have already been proven to be not practical or even not completely secure [31, 64, 73]. That is why we do not discuss them in this work. As well as the linearly homomorphic signature schemes defined over bilinear groups (see

Sect. 4.1), also the construction of the RSA-based linearly homomorphic signature schemes exploits the homomorphic property of DLP. In the following, the existing RSA-based solutions are presented and compared.

4.2.1   Secure Network Coding Over the Integers, by Gennaro et al. [36]

In 2010, Gennaro et al. presented in [36] a Standard-RSA-based signature scheme. In order to have a concrete example, the authors of [50] implemented and tested the scheme in Linux. Running, for instance, the 512-bit RSA signature scheme proposed in [36] with an exponent of 1024 bits took 3.2 ms. Using a 112-bit elliptic curve the same exponentiation could be performed in 7.79 ms. This shows that the algorithms of the scheme proposed by Gennaro et al. can be run in reasonable time. In fact in this scheme the linear combinations involve only 8-bit long coefficients and the length of the signature depends on the integer n chosen in the RSA assumption. It is proven in the Random Oracle Model that this scheme is unforgeable against the weak adversary. The level of privacy guaranteed by the scheme has not been specified yet.

4.2.2   Adaptive Pseudo-Free Groups and Applications, by Catalano et al. [21]

The scheme proposed by Catalano et al. in [21] relies on the Strong-RSA assumption and the authors proved that all signatures built on RSA are instances of their general framework. The signature is composed of two λ-bits long integers, the message to be signed and the file identifier τ, where λ is the security parameter. The verification performs an exponentiation with a λ-bit integer and one multi-exponentiation with an exponent of λ bits (see [22]). The scheme is proven to be secure in the Standard Model against the weak adversary. Also in this case, the proof in the Standard Model can be modified such that it is secure against the strong adversary. The privacy level still has to be clarified.

Remark 4.1. The linear combinations are performed over the integers. This leads to the fact that the number of admissible linear combinations is bounded, otherwise the vector coordinates would grow too large.

4.2.3   Efficient Network Coding Signatures in the Standard Model, by Catalano et al. [22]

The scheme described in [22] was presented by the same authors of [21] 1 year later. This new one works over $\mathbb{Z}_M$, where $M = pq$ and p, q are two safe primes. It relies on the same hardness assumption (Strong-RSA), but it is an improvement in terms of size. The signature contains only one element of $\mathbb{Z}_M$ and one integer of λ bits, where λ is the security parameter. The verification needs an exponentiation·with a λ-bit integer and one multi-exponentiation with λ-bits exponents. In addition, the number of admissible linear combinations is not bounded anymore like in [21]. Indeed they are computed modulo a certain prime number such that the vector coefficients cannot grow beyond it. The scheme is proven secure in the Standard model and can cope with the weak adversary. However, the privacy level has not been analyzed yet.

Remark 4.2. Another important feature of this scheme is that the length of the vector to be signed is unbounded.

4.2.4   Improved Security for Linearly Homomorphic Signatures: A Generic Framework, by Freeman [34]

Besides the schemes described in Sect. 4.1.5, the generic framework presented in [34] also shows how to adapt two RSA-based signature schemes such that they support linear operations. More precisely, the scheme proposed in [35] relying on the Strong RSA assumption and the scheme proposed in [44] relying on the Standard RSA assumption are transformed. Like the schemes based on bilinear groups also these two signature schemes achieve a high security level. The unforgeability is proven secure in the Standard Model against the stronger adversary and the signatures are weakly context hiding.

Remark 4.3. While the signature scheme presented in Sect. 4.2.3 is unbounded regarding the length of the vector to be signed. This this not true for the two constructions presented in this paper.

4.2.5   Summary of RSA-Based Linearly Homomorphic Signature Schemes

In Table 4.2 the properties of the RSA-based linearly homomorphic signatures discussed so far are summarized. We see that there is only one scheme, the one discussed in Sect. 4.2.4, that is secure against the strong adversary and provides weak privacy. If facing a weak adversary is sufficient also the schemes described in

Table 4.2 RSA-based linearly homomorphic signature schemes

Signatures	Hard. Ass	Adversary	Privacy	Model	Efficiency
Section 4.2.1	Standard-RSA	Weak	Ø	ROM	
Section 4.2.2	Strong-RSA	Weak	Ø	Standard	
Section 4.2.3	Strong-RSA	Weak	Ø	Standard	Improves Sect. 4.2.2[a]
Section 4.2.4	Strong-/Standard-RSA	Strong	Weak	Standard	

[a]With respect to the length of the signature

Sects. 4.2.1–4.2.3 can be used. Note that the advantage of the latter scheme is that it is proven secure in the Standard Model and that it is not bounded regarding the length of the vector to be signed nor with respect to the number of admissible linear combinations to be performed.

4.3 Lattice-Based Linearly Homomorphic Signature Schemes

In this section we describe the linearly homomorphic signature schemes whose hardness assumption is defined over the lattices. Lattice-based primitives are expected to be resilient against attacks by quantum computers. The construction of lattice-based linearly homomorphic signature schemes comes from the following fact. A vector space V is given by a basis of vectors $v_1, v_2, \ldots, v_n \in \mathbb{F}^n$. To sign the vector space V, first, the individual vectors $v_1, v_2, \ldots, v_n$ are signed. A signature σ_i for vector v_i is a low-norm vector in $\mathbb{Z}^m$ for an integer m such that the following property is satisfied:

$$\mathbf{A}_V \cdot \sigma_i = q \cdot v_i \mod 2q, \tag{4.1}$$

where q is an odd prime and $\mathbf{A}_v \in \mathbb{Z}_{2q}^{m \times n}$ is a matrix. The homomorphic property is given, because the above property still holds for a signature $\sigma := \sum_{i=1}^{n} \sigma_i$ on the vector $v := \sum_{i=1}^{n} v_i$. Later it has been shown that even constructions providing polynomial functions and fully homomorphic solutions can be developed using lattices. In this section we focus on the linearly homomorphic signature schemes. The constructions supporting polynomial and arbitrary functions are instead presented in Sects. 4.4 and 4.5, respectively.

4.3.1   Linearly Homomorphic Signatures over Binary Fields and New Tools for Lattice-Based Signatures, by Boneh and Freeman [15]

The signature scheme described in [15] is the first one that is assumed to be resistant against quantum attacks. The hardness assumption exploited is k-SIS. Referring again to the notation used in Sect. 3.1.3, the public key is $m + m \log q$ bits long, while the length of the signature is $2m + 2m \log q$ bits. The space overhead needed to run the verification algorithm is of $m^2 + 2m^2 \log q + m^2 (\log q)^2$ bits (see [45]). Note that this signature scheme is built on the non-homomorphic one presented in [38]. The scheme is resistant to the weak adversary and proven secure in the Random Oracle Model. Furthermore, it is weakly context hiding in terms of privacy. Note that if this property is provided the number of linear operations that can be authenticated by the scheme is bounded (see Remark 4.4).

4.3.2   Lattice-Based Linearly Homomorphic Signature Scheme over Binary Fields, by Wang et al. [65]

The work presented in [65] by Wang et al. relies on the SIS hardness assumption. In this scheme, independent of the security parameter, the length of the public key and the signature is $m \log q$ bits. The space overhead consumed during the verification is $m^2 (\log q)^2$ (see [45]). As pointed out by the authors of [65] and proven in the Random Oracle Model, the scheme is secure against the weak adversary. Furthermore, it is weakly context hiding.

Remark 4.4. As discussed in [45] there is an open problem which concerns all the current lattice-based homomorphic signature schemes over binary field. That is, they are L-limited, where L is the upper-bound on the maximum number of signatures that can be combined.

4.3.3   Summary of Lattice-Based Linearly Homomorphic Signature Schemes

In Table 4.3 the properties of the linearly homomorphic signatures discussed so far are summarized. We see that only solutions that can cope with the weak adversary, provide weak privacy, and are proven secure the Random Oracle Model are available. From these two signature schemes the one presented in Sect. 4.3.2 improves the scheme presented in Sect. 4.3.1 with respect to the public key and signature length and space overhead occupied during the verification procedure.

Table 4.3 Lattice-based linearly homomorphic signature schemes

Signatures	Hard. Ass	Adversary	Privacy	Model	Efficiency
Section 4.3.1	k-SIS	Weak	Weak	ROM	
Section 4.3.2	SIS	Weak	Weak	ROM	Improves Sect. 4.3.1[a]

[a]With respect to the public key and signature length and the space overhead needed during the verification

4.4 Homomorphic Signature Schemes for Polynomial Functions

Gentry [37] used lattice-based cryptography to build the first encryption scheme supporting both additions and multiplications on ciphertexts. On the basis of this achievement also corresponding homomorphic signature schemes were developed supporting evaluations of multivariate, bounded-degree polynomials on authenticated data. These schemes are presented in the following sections.

4.4.1 Homomorphic Signatures for Polynomial Functions, by Boneh and Freeman [14]

The signature scheme proposed in [14] is the first homomorphic signature scheme for polynomial functions. It relies on the SIS hardness assumption and is defined over ideal lattices. Furthermore, it fulfills the definition of succinctness of a signature's length, as defined in Sect. 3.2. In fact the signature depends logarithmically on the data size. The verification algorithm involves an upper-bound check on the norm of the signature, two polynomial evaluations with modulo, and a hash function evaluation. The scheme can cope with the weak adversary. The security is proven in the Random Oracle Model. The level of privacy achieved by the signature scheme still needs to be clarified.

4.4.2 Homomorphic Signatures for Polynomial Functions with Shorter Signatures, by Hiromasa et al. [43]

Hiromasa et al. presented in [43] a signature scheme that is very similar to the one proposed in Sect. 4.4.1, i.e. it also relies on the hardness assumption SIS. The interesting property is that this scheme generates shorter signatures. This can be achieved thanks to the employment of an alternative algorithm in the signing process. However, this is done at the expense of having a longer secret key.

Regarding all the other properties and parameters, the signature scheme is proven secure against the weak adversary in the Random Oracle Model. Privacy is still a non specified property.

4.4.3 Homomorphic Signatures with Efficient Verification for Polynomial Functions, by Catalano et al. [23]

The homomorphic signature scheme described in [23] relies on the k-Augmented Power Multilinear Diffie-Hellman Problem (k-APMDHP). That is a hardness assumption that the authors defined by themselves and we refer directly to [23] for its definition. This scheme is claimed by the authors to outperform the signature scheme introduced by Boneh and Freeman in [14] (and also the one described in [43]) in the following terms.

1. The scheme is proven secure in the Standard Model instead of the Random Oracle Model.
2. The adversary is not assumed to query signatures on messages in a given data set all at once. That is, the scheme is proven to be secure also in the presence of a strong adversary.
3. The scheme achieves better performances in terms of efficiency as the verification of a signature with respect to a polynomial function f takes less time then computing the function itself. That is, the scheme is efficient in this amortized sense, since the verification algorithm is supposed to be run many times, while the signing algorithm is performed just once.

However, the scheme relies on multilinear maps and it is still a work-in-progress to define a practical and efficient one (see [27]). Thus, this homomorphic signature scheme might not be employed until a good multilinear map is designed. Moreover, the open problem stated by the authors in [14] of building a homomorphic signature scheme providing privacy has still not been solved.

4.4.4 Summary of Homomorphic Signature Schemes for Polynomial Functions

The properties of the signature schemes supporting polynomial functions are summarized in Table 4.4. We see that none of the three schemes available are at least weakly-context hiding. Furthermore, as already mentioned, the only scheme that is secure against the strong adversary, the signature scheme presented in Sect. 4.4.3, relies on multilinear maps which is still work in progress. The other two schemes can only cope with the weak adversary and are proven secure in the Random Oracle Model only.

Table 4.4 Homomorphic signature schemes for polynomial functions

Signatures	Hard. Ass	Adversary	Privacy	Model	Efficiency
Section 4.4.1	SIS	Weak	∅	ROM	
Section 4.4.2	SIS	Weak	∅	ROM	Improves Sect. 4.4.1[a]
Section 4.4.3	k-APMDHP	Strong	∅	Standard	Improves Sect. 4.4.2[b]

[a] With respect to the signature length and the expensive of having a larger secret key
[b] With respect to the running time of the verification algorithm

4.5   Fully Homomorphic Signature Schemes

After Gentry provided the first fully homomorphic encryption scheme further improvements have been made. Later constructions, for instance, allow to construct an identity-based FHE scheme, make bootstrapping optional, eliminate the modulus switching, etc. In line of these improvements, also new homomorphic signature schemes supporting arbitrary functions rather than merely polynomial ones have been developed. In Sect. 4.5.1 a construction using homomorphic techniques from [39] is presented. This approach has been further improved and the resulting signature scheme is presented in Sect. 4.5.2. Another approach exploits the fact that any function can be transformed into a circuit consisting of NAND gates. More precisely, the signature scheme contains an algorithm computing the signed output of such NAND gate after receiving two signed inputs. Consequently, this scheme allows to evaluate any function represented by a NAND gate circuit. This work bases on the *HTDF* functions proposed by Gorbunov et al. [41] and is described in Sect. 4.5.3.

4.5.1   Leveled Fully Homomorphic Signatures from Standard Lattices, by Gorbunov et al. [41]

In [41], the first fully homomorphic signature scheme is proposed. This scheme bases on the hardness assumption SIS and can evaluate arbitrary circuits over signed data. With respect to efficiency, the costs for verification is as high as computing the function f. The signature scheme can be proven secure either in the Random Oracle Model or in the Standard Model. The former case allows for short public parameters, while in the latter case they are even longer than the total size of the data set.

Note that, in both cases, instead, the signature's size is independent of the data size and of the circuit size. Though, this does not mean that the signature is short: it is indeed dependent of the depth d of the circuit, which is an *a priori* fixed parameter. Therefore, even though we can in principle perform any kind of transformation on the authenticated data, this is done at the expense of having a larger and larger signature.

The adversary these schemes can cope with is the weak one. A particular technique is available [13] that allows to convert the schemes such that they can cope with the strong adversary. However, as pointed out by [19] this might lead to a quite inefficient scheme, since only few and short messages can be signed. Furthermore, they are claimed not to lack information about the original data beyond the outcome of the transformation itself. Therefore, according to the terminology adopted in this work, weakly context hiding privacy is provided.

4.5.2 Adaptively Secure Fully Homomorphic Signatures Based on Lattices, by Boyen et al. [19]

The work presented in [19] is the second proposal for a fully homomorphic signature scheme. It is still based on lattices, assumed to provide security even in the presence of quantum computer. This paper can be thought as a concurrent work to the scheme presented in Sect. 4.5.1. Indeed some improvements have been done, though arising some problems not present in the aforementioned paper. Specifically, the scheme cannot sign arbitrary circuits any more: rather the ones with poly-logarithmic depth or the ones with polynomial depth. In the first case the hardness assumption is SIS, while in the second case the scheme relies on sub-exponential SIS. On the other hand, the efficiency is claimed to be definitely improved even though there is no discussions about the signature's size. More precisely, the verification process involves the computation of the norm of a matrix, two matrix multiplications, one matrix addition, and a modulo operation. An important improvement of this work is that the scheme can cope with the strong adversary and this is proven in the Standard Model. Unfortunately none of the possible level of privacy is achieved, making the protocol not applicable for many real-life applications.

4.5.3 Leveled Strongly-Unforgeable Identity-Based Fully Homomorphic Signatures, by Wang et al. [66]

The signature scheme presented by Wang et al. in [66] is the first fully homomorphic signature scheme that is strongly unforgeable, see Sect. 1.2.3. The hardness assumption employed is SIS. This work is claimed to extend the signature scheme described in Sect. 4.5.1 in terms of the trapdoor function used. In fact, this signature scheme relies on a trapdoor function that is also identity-based. This means that the verification algorithm requires only the public parameters and the file identifier (i.e. the tag), leading to an easier key management. In addition, the scheme is secure against the weak adversary and the security proof is set in the Random Oracle Model. Privacy needs to be clarified.

Table 4.5 Fully homomorphic signature schemes

Signatures	Hard. Ass	Adversary	Privacy	Model	Note	Efficiency
Section 4.5.1	SIS	Weak	Weak	Standard	Large param.	
Section 4.5.1	SIS	Weak	Weak	ROM	Short param.	
Section 4.5.2	SIS	Strong	None	Standard	Poly-log depth	Improves Sect. 4.5.1[a]
Section 4.5.3	SIS	Weak	None	ROM		

[a]Claimed by the authors of the scheme described in Sec. 4.5.2. Further analysis are needed

4.5.4 Summary of Fully Homomorphic Signature Schemes

In short, we summarize the above schemes in Table 4.5. The table shows that there is only one scheme, the one presented in Sect. 4.5.2, that can cope with the strong adversary. However, for this approach privacy still needs to be clarified. In case weak privacy is required and security against the weak adversary is sufficient the schemes described in Sect. 4.5.1 can be used. One is proven secure in the Random Oracle Model only while the other one is proven secure in the Standard Model at the expenses of large parameters. If security against the weak adversary is sufficient and privacy is not required also the scheme presented in Sect. 4.7.1 is a good choice as it allows for an easier key management.

4.6 Multiple Sources Linearly Homomorphic Signature Schemes

In the following, the existing signature schemes supporting linear operations over signed messages generated by different users are presented. All these approaches exploit that constructions build on bilinear maps have homomorphic properties. However, the difference compared to the signature schemes presented in Sect. 4.1 is that they support linear operations even on messages that have been signed by different users holding different secret keys.

4.6.1 Signatures for Multi-Source Network Coding, by Czap and Vajda [30]

The scheme presented in [30] adapts the linearly homomorphic signature scheme introduced by Boneh et al. in [18] (see Sect. 4.1.1) to the multiple sources case. The properties provided by the two schemes are similar. This multiple sources linearly homomorphic signature scheme relies on co-CDH and it is proven secure in the Random Oracle Model against the weak adversary. In the verification procedure there are $N + 1$ asymmetric pairings, where N is the number of the messages to be combined, i.e. the number of users involved. Furthermore, the signature is a vector of N entries, as each entry $i \in \{1, 2, \ldots, N\}$ is the corresponding signature for the message m_i. Privacy needs to be clarified.

4.6.2   Short Signature Scheme for Multi-Source Network Coding, by Yan et al. [69]

The multiple sources linearly homomorphic signature scheme introduced in [69] relies on the hardness assumption co-CDH and the pairings are asymmetric. As for the scheme described in Sect. 4.6.1, it is proven secure in the Random Oracle Model and can cope with the weak adversary, since an attacker can query a data set once. With respect to the previous scheme, this achieves better performances in terms of efficiency: the verification algorithm needs one pairing and the signature is as short as the shortest signature in the respective single-user scheme. In addition the authors provide an analysis showing that the communication overhead needs less bits than the scheme presented in Sect. 4.6.1. Privacy has not being clarified yet.

4.6.3   Efficient Multiple Sources Network Coding Signature in the Standard Model, by Zhang et al. [75]

The scheme presented in [75] and in [62] is the first multiple sources homomorphic signature scheme proven secure in the Standard Model. The hardness assumption the scheme relies on is q-SDH and it is built adapting the linearly homomorphic signature schemes described in Sect. 4.1.1 and in Sect. 4.1.4. If the linear computations are performed over N messages, then the signature is composed of one group element, N signatures and one integer. The verification algorithm needs $N + 1$ asymmetric pairings. However, the scheme copes with a weaker attacker then the weak adversary discussed so far. The privacy level achieved needs to be discussed.

Remark 4.5. The authors of [75] claim that the scheme really supports multiple sources, as any source can transmit a whole file and not just a part of it, like for the scheme introduced in Sect. 4.6.2.

4.6.4   Summary of Multiple Sources Linearly Homomorphic Signature Schemes

The properties discussed so far about the existing multiple sources linearly homomorphic signature schemes are provided in short in Table 4.6. The table shows that

Table 4.6 Multiple sources linearly homomorphic signature schemes

Signatures	Hard. Ass	Adversary	Privacy	Model	Efficiency	Pairing
Section 4.6.1	co-CDH	Weak	$\emptyset$	ROM	$N + 1$ Pairings	Asymm.
Section 4.6.2	co-CDH	Weak	$\emptyset$	ROM	1 Pairing	Asymm.
Section 4.6.3	q-SDH	–	$\emptyset$	Standard	$N + 1$ Pairings	Asymm.

there is no scheme that provides privacy or can cope with the strong adversary. The schemes presented in Sects. 4.6.1 and 4.6.2 are secure against the weak adversary and has been proven secure in the Standard Model. The scheme presented in 4.6.3 can only cope with an adversary that has to be even less powerful than the weak adversary.

4.7 Linearly Homomorphic Aggregate Signature Schemes

In many practical applications, it might be necessary not only to perform computations on signatures generated by different users, but also to aggregate them. The merely (multiple sources) linearly homomorphic signature schemes do not fulfill this additional task. Up to our knowledge, there are only two signatures which are both homomorphic and aggregative. Both of them are lattice-based and are therefore assumed to be secure even in the presence of quantum computers. In this section further information about these constructions and their properties is provided.

4.7.1 A Homomorphic Aggregate Signature Scheme Based on Lattice, by Zhang et al. [74]

In [74] the first linearly homomorphic aggregate signature scheme is proposed. This scheme relies on the hardness assumption ISIS, that is defined over lattices. Without taking into account the security parameter, the verification algorithm needs $4m^2 (\log q)^2$ bits in terms of space overhead. Furthermore, the length of the signature is $2m \log q$ bits, while the length of the public key is $m \log q$ bits. With respect to security, the scheme is proven secure in the Random Oracle Model against the strong adversary. However, privacy is not clarified yet.

4.7.2 An Efficient Homomorphic Aggregate Signature Scheme Based on Lattice, by Jing [45]

The linearly homomorphic aggregate signature scheme described in [45] is defined over the latices, as it relies on the hardness assumption SIS. Regarding the space overhead, the verification algorithm uses $m^2(\log q)^2$ bits (also in this case the security parameter is not taken into account). The length of the signature is $m \log q$ bits. The length of the public key is $m \log q$ bits, like for the public key of the scheme described in Sect. 4.7.1. The security proof is in the Random Oracle Model and the scheme is secure against the strong adversary. In terms of privacy, it is weakly context hiding. Note that this scheme is a variant of the linearly homomorphic signature introduced in [15], where the same privacy property holds and also apply to the multi-users case.

4.7.3 Summary of Linearly Homomorphic Aggregate Signature Schemes

The properties of the two existing linearly homomorphic aggregate signature schemes discussed are reported in Table 4.7. Both signature schemes available are secure against the strong adversary. In particular the scheme presented in Sect. 4.7.2 is even context hiding and improves the one presented in Sect. 4.7.1 with respect to efficiency.

Table 4.7 Linearly homomorphic aggregate signature schemes

Signatures	Hard. Ass	Adversary	Privacy	Model	Efficiency
Section 4.7.1	ISIS	Strong	Ø	ROM	
Section 4.7.2	SIS	Strong	Weak	ROM	Improves Sect. 4.7.1[a]

[a]With respect to the overhead of the verification algorithm and the length of the signature

Chapter 5
Suitable Homomorphic Signature Schemes for eVoting, Smart Grids, and eHealth

Abstract The signature schemes presented in Chap. 4 are discussed from an abstract and very general point of view. In this chapter the requirements a scheme needs to provide to be applied for a certain application will be highlighted. Specifically, in this section electronic voting, smart grids, and electronic health records are discussed. Each of the following sections is dedicated to one of them. After a brief description of the use case in question, the requirements for a homomorphic signature scheme are discussed, the state of the art is presented, and possible future work is highlighted.

5.1 Electronic Voting

Since the existence of democracy, several voting schemes have been designed allowing people to express their opinion. In order to be *general*, *direct*, *free*, *equal*, and *secret* an election needs to fulfill several security requirements. These include, among others, correctness. More precisely, correctness requires that only votes cast by eligible voters are tallied and that the election outcome is computed correctly without removing and/or adding ballots.

Paper based voting schemes are currently the most widely used ones. Though, they have the drawback that only people present during the tallying procedure can verify that the votes cast are counted correctly.

There are several electronic Voting (eVoting) schemes that support a remotely verifiable tallying process using a public bulletin board. More precisely, during the vote casting process each voter receives a receipt containing some information, in most schemes the own vote in encoded form. After the polls closed all encoded votes contained in the ballot box are published on the bulletin board and each voter can verify that the own vote has been recorded as cast using his/her receipt (*individual verifiability*). In addition, during the tallying process some audit data is published that allows anybody to check that all votes recorded have been tallied correctly (*universal verifiability*). If votes are cast in encrypted form there are mainly two approaches how they can be processed. First, the ciphertexts are made anonymous, e.g. using a mix-net [59], followed by decryption and tallying. Second, the votes are tallied in encrypted form and are decrypted afterwards. Recently, a

© The Author(s) 2016
G. Traverso et al., *Homomorphic Signature Schemes*, SpringerBriefs in Computer
Science, DOI 10.1007/978-3-319-32115-8_5

rerandomizable signature scheme for the first approach has been proposed in [28]. The homomorphic signatures schemes described in this survey address the second, more efficient, tallying procedure.

Employing homomorphic signatures would allow to verifiably tally votes that have been authenticated leading to an authenticated result. Note that here two scenarios need to be distinguished. In the first scenario all votes are signed using the same global, but secret, election signing key. In the second scenario several secret keys are used to sign individual votes or set of votes. An example for the first scenario are poll-site voting schemes where the voters cast their vote in a polling station using official election hardware, e.g. by casting votes using voting machines or by scanning filled out ballots. Here a possible application for homomorphic signatures is that the device in question signs the digitally recorded ballots using a global election signing key. Thus, all votes published on the bulletin board are authenticated and an election outcome signed with this election key can be computed. The second scenario is interesting for online voting schemes where the votes are cast remotely by the individual voters. Here each voter submits his/her own vote signed with his/her individual signing key. In this case tallying the votes published on the bulletin board leads to an election outcome that is signed by all voters. While the first scenario requires regular homomorphic signatures, in the second scenario homomorphic signatures for the multi-users scenario are needed. In both cases, the operation performed over the signed votes range from simple additions of plaintext votes (see [25]), to polynomial functions (see [2]), e.g. simple operations on encrypted data, up to arbitrary functions (see [68]). Thus, for this use case, linearly, polynomial, fully, and homomorphic signatures for the multi-users case are of interest.

In order to be employable in the electronic voting scenario the homomorphic signature schemes have to fulfill the following requirements. With respect to the hardness assumption, it is sufficient that the signature schemes used are based on the classical problems of Integer Factorization and Discrete Logarithm Problem. Indeed, the election result is made public as soon as the counting is completed. Therefore, there is no need for long-term protection of authenticity. Although it is preferable to determine the election outcome as fast as possible, efficiency is in general a less critical aspect for the tallying process. Indeed, computationally powerful devices (e.g. laptops, voting machines) are usually employed. For the same reasons, having a succinct signature is also less important. In many schemes the signed votes are not published before the poll is closed and the voters do not receive any feedback whether the signatures are correct or not. This is for instance the case when manipulated hardware is used. In this case it would not be possible for an adversary to submit a second set of signed votes when it could not forge the first set of signatures successfully. The hardware would be replaced and the voting process would be repeated. Therefore, it is sufficient for the scheme to be secure against the weak adversary. However, there might be other schemes where the signature to the data cast is verified directly and feedback is given to the signer. In this case the signature scheme must be secure against the strong adversary. With respect to privacy, it is sufficient for a homomorphic signature scheme to achieve the weak context hiding level. In fact it is well known that the set of possible messages are votes and that the admissible functions correspond to the election methods. If only encrypted votes are signed, then privacy is not needed at all.

The use case of electronic voting comes with quite low requirements. The signature schemes need not to be post-quantum secure, efficiency is not a critical aspect, the schemes does only have to cope with the weak adversary, and if only encrypted data is signed no privacy is needed. Thus, actually all schemes presented in Sects. 4.1–4.3 are suitable for this application. Another criteria to select one of the valid signature schemes can be their efficiency. For this a rigorous efficiency evaluation is needed. Note that using the linearly homomorphic signature schemes only linear operations can be performed on the ciphertexts. If a homomorphic encryption scheme is used where the ciphertexts have to be multiplied, such as El Gamal [33] or Paillier [58], homomorphic signature schemes supporting polynomial functions are needed. In case the voting system processes the votes in plain the signature schemes must provide at least weak privacy. However, also for this scenario plenty of signature schemes are available.

With respect to the schemes supporting polynomial functions, none of the approaches provide weak privacy. Therefore, none of them can be considered in case the votes are not encrypted. If only encrypted votes are processed it should be noted that depending on the function performed this can lead to a polynomial with a very high degree. One example is the case where all encrypted votes cast are multiplied which is common practice in eVoting. Using homomorphic signatures for polynomial functions in this scenario might lead to a very inefficient tallying procedure.

For the eVoting use case there are several fully homomorphic signature schemes available that meet the minimal requirements. There is even a scheme that is secure against the strong adversary and can be used for eVoting schemes where the signature is directly verified and feedback is given to the sender. In addition there are signature schemes available that provide weak privacy in case the votes are processed in plain. The same holds true for homomorphic multiple sources signature schemes. However, all signature schemes for the multi-users scenario only support linear functions. Thus, it would be desirable to develop schemes that also provide more complex operations.

In short one can say that for voting schemes with a simple election method (the addition of votes) there are signature schemes available that provide reasonable efficiency. Although solutions for more complex operations and the multi-users scenario have been proposed it is not clear how they perform when a thousand or even a million votes have to be counted. Thus, before one of these approaches is used, an efficiency analysis have to be performed. Note that although efficiency is not a very critical property for this use case, still during elections a huge amount of data needs to be processed and the election outcome should be available in reasonable time.

5.2 Smart Grids

Smart grids allow to introduce intelligent electricity generation, load balancing, resource allocation, and dynamic pricing on the basis of real-time power consumptions. The drawback of this new technique is that the collected data allows to generate profiles of the energy consumers. Thus, in order to preserve data privacy of individual households, the so called in-network aggregation is performed. More precisely, the measured data is routed through a set of smart meters where each smart meter aggregates its input. Thus, the result reported to the supplier only provides information about a district but hides the fine-grained individual metering data. To prevent that the meters on the route can see the intermediate results the measurements are encrypted by the smart meters using a homomorphic encryption scheme and aggregated using the homomorphic property. Besides homomorphic encryption also digital signatures provide important functionalities for this use case. They protect against unintentional errors and prevent adversaries from altering messages. However, to be compatible with privacy-preserving in-network data aggregation, signatures with homomorphic properties are needed. They can be used to sign the encrypted metering data and be aggregated along with the corresponding ciphertexts at each intermediate node [51, 70]. This allows the energy supplier to verify the correctness of the aggregation by checking the consistency between the aggregated result and the aggregated signatures.

As we already mentioned, in the framework of smart grids only encrypted data are signed. If each smart meter on the aggregation route makes use of the same secret key in the signing procedure, then linearly homomorphic signature schemes and homomorphic signature schemes for polynomial functions are taken into account. More precisely, if in the employed homomorphic encryption scheme the ciphertexts are added, then a linearly homomorphic signature schemes is needed. If instead the ciphertexts are multiplied, then a homomorphic signature scheme supporting polynomial functions has to be employed. On the other hand, if the secret keys differ for each smart meter, then multiple sources homomorphic signatures are the suitable ones.

In this context, it is sufficient for a homomorphic signature scheme to be based either on the Discrete Logarithm Problem, or on the Integer Factorization Problem. Indeed there is no need for a long-term storage of the consumptions. However, the signature schemes should provide high performances in terms of efficiency. In fact the power consumption needs to be reported in real-time and therefore signature generation must be fast. In addition, the data are computed and aggregated by the smart meters and they have only restricted resources. Thus, the signatures' size should be as short as possible. The homomorphic signature schemes should also cope with the strong adversary. Indeed, since there is the possibility to resent rejected data, the attacker can perform queries multiple times. For smart metering, privacy is not an issue since only encrypted data are signed. Thus, there are no constraints regarding the context hiding level.

In this use case the requirements are more restrictive than for the electronic voting use case. The signature scheme used must be secure against the strong adversary and also efficiency is an important criteria. Consequently, there are some signature schemes that cannot be taken into account for this application. One example are linearly homomorphic signature schemes that are based on lattices, because they are not secure against the strong adversary. However, there are schemes available that are based on bilinear groups or RSA, which is sufficient since the signatures must not be post-quantum secure. However, the linearly homomorphic signature schemes can only be used if the ciphertexts generated are simply added during the aggregation. If this is not the case homomorphic signature schemes supporting polynomial functions or fully homomorphic signature schemes are needed. For the polynomial case the only scheme that is secure against the strong adversary requires the existence of efficient multilinear maps and developing such maps is still work in progress. However, there are fully homomorphic signature schemes and those for the multi-users case available. Nevertheless, also for the smart grid use case homomorphic signature schemes for the multi-users scenario that support polynomial functions or that are fully homomorphic are of interest. Furthermore, efficiency is a very important criteria and it is not clear whether the schemes available provide a reasonable level of efficiency. Thus, also for this use case an efficiency analysis would be valuable.

5.3   Electronic Health Records

In the last years, there has been an increasing interest in moving to digital health records. Indeed in many European countries recording such data electronically is part of the national health informatics strategies (see [47]) and United Kingdom is one of the most advanced in this respect (see [61]). Also in the United States the adoption of electronic health records is becoming more and more widespread [10]. Recording health information in a digital fashion make it more reliable and easier to access by different medical facilities, such as medical practices, hospitals, health insurances, medical institutes, and pharmacies. The data stored can be used for merely consultations, but not only for that: one may want to perform computations over them, such as statistical calculations. In case the data are authenticated by a homomorphic signature scheme, then the above issue can easily be addressed, as discussed in [54]. In fact, let us suppose that a doctor has signed several data regarding its patients. Then, another institution, e.g. a medical institute, can perform a computation on a specific data set, e.g. measured blood pressures, outputting the final result already authenticated accordingly. This scenario can be extended to an input set signed by several doctors in a hospital, where health records are stored in a common data base, and even to several hospitals. In this case, homomorphic aggregate signatures are required to perform the computations, since the original data are signed by doctors using different secret keys.

Summarizing, in this context linearly homomorphic signatures, homomorphic signatures for polynomial functions, fully homomorphic signatures, and homomorphic aggregate signatures are of interest. In addition, depending on for how long a certain data is stored, the schemes to take into account can be either the ones based on classic problems (Integer Factorization and Discrete Logarithm Problem) or the ones based on lattices problems. Due to the sensitive nature of the information involved, the properties that a homomorphic signature scheme has to satisfy are demanding. Indeed, the efficiency of the computation should be high: a lot of information is stored every day and the function to be applied might be expensive. However, it is assumed that the devices employed in this scenario are not computationally weak. Therefore, also a large signature's size can still be accepted in case it is not succinct. The scheme should be safe against the strong adversary, since there is the possibility to repeat queries multiple times. In addition, due to the sensitivity of the information, privacy is an important property. However, the weakly context hiding level of privacy is sufficient: a homomorphic signature scheme achieving such level of privacy already does not leak information about the original data set. That is, no one can see for example the real blood pressure values of the patients, and only the result of the computations is revealed. If the data is encrypted then privacy is not a requirement.

The requirements for the eHealth use case are quite similar to those for smart grids. The schemes used should be secure against the strong adversary and provide weak privacy if data is signed and processed in plain. Also efficiency is an important criteria although this is less critical compared to smart grids. In fact in the eHealth scenario usually the devices employed are more powerful. Nevertheless, also for this use case performing an efficiency analysis and developing homomorphic signature schemes for the multi-users scenario supporting less restrictive functions would be desirable. Furthermore, health records might be stored for several decades. Thus, it would be interesting to see whether a signature scheme can be developed that can cope with the strong adversary, provides weak privacy, and is post-quantum secure.

Chapter 6
Conclusion

In this work a formal definition of the general framework regarding homomorphic signature schemes is provided. Starting from such framework, it is also shown how linearly homomorphic signature schemes, homomorphic signature schemes for polynomial functions, fully homomorphic signature schemes, and homomorphic aggregate signature schemes are derived. Afterwards, the first up-to-dated survey about all the currently existing homomorphic signature schemes is provided, where each scheme is singularly described and analyzed. In addition three interesting use cases for homomorphic signature schemes were presented. These are: electronic voting, smart grids, and electronic health records. For each use case the minimal requirements a suitable homomorphic signature scheme should fulfill are discussed and the existing homomorphic signature schemes that properly address these requirements are presented. Based on these observations, directions for future work are suggested that would address the faults of the current state of the art.

One of the most important directions for future research with respect to homomorphic signature schemes is efficiency. So far, only partial comparisons have been provided, proposing a qualitative rather then quantitative description in this regard. However, a deep analysis involving all the existing schemes is not available yet, even though it would be a very valuable contribution. Indeed, having a clear insight about efficiency, would provide a better understanding of the schemes' performances in practice. Furthermore, it would show how these schemes behave in real-life situations, such as when they have to run on computationally weak devices or process large amounts of data.

With respect to linearly homomorphic schemes, it would be useful to design approaches providing both a good efficiency level and safety against the strong adversary. Such schemes would be suitable to address the minimal requirements of smart grids. In addition, for electronic health records it would be desirable that linearly homomorphic signature schemes are developed that are expected to be resilient even against quantum computer attacks and that achieve at least weak privacy.

© The Author(s) 2016
G. Traverso et al., *Homomorphic Signature Schemes*, SpringerBriefs in Computer
Science, DOI 10.1007/978-3-319-32115-8_6

Regarding homomorphic signature schemes supporting polynomial functions, it would be interesting to design a scheme that provides at least weak privacy, so that it can be used in the context of electronic voting. Together with privacy, in order to be applicable for electronic health records, homomorphic signature schemes for polynomial functions need to be developed that are secure against the strong adversary.

With respect to the fully homomorphic signature schemes, an important topic for future work is to build a scheme that copes with the strong adversary and achieves at least weak privacy. This would allow to use it together with electronic health records.

The state of the art with respect to homomorphic multiple sources signature schemes is that only linear functions can be provided. Thus, for future work it would be interesting to see whether signature schemes for the multiple-users case can be developed that are less restrictive with respect to the operations supported.

Finally, in this work we looked at the use cases only from a high level point of view. Thus, further research should be done before using homomorphic signature schemes for the mentioned applications. However, the discussion of the use cases also showed that not for each scenario an appropriate signature scheme is available and that extensive efficiency analyses are missing. Furthermore, the introduction of homomorphic signature schemes to several use cases needs further research. In our perspective these are interesting directions for future work and we plan to work on these matters in the future.

References

1. Abe M, Haralambiev K, Ohkubo M (2010) Signing on elements in bilinear groups for modular protocol design. IACR Cryptology ePrint Archive, 2010:133
2. Adida B, Rivest RL (2006) Scratch & vote: self-contained paper-based cryptographic voting. In: Proceedings of the 5th ACM workshop on privacy in electronic society. ACM, New York, pp 29–40
3. Agrawal S, Boneh D, Boyen X, Freeman DM (2010) Preventing pollution attacks in multi-source network coding. In: Public key cryptography–PKC 2010. Springer, Berlin, pp 161–176
4. Ahn JH, Boneh D, Camenisch J, Hohenberger S, Waters B et al (2012) Computing on authenticated data. In: Theory of cryptography. Springer, Berlin, pp 1–20
5. Attrapadung N, Libert B (2011) Homomorphic network coding signatures in the standard model. In: Public key cryptography–PKC 2011. Springer, Berlin, pp 17–34
6. Attrapadung N, Libert B, Peters T (2012) Computing on authenticated data: new privacy definitions and constructions. In: Advances in cryptology–ASIACRYPT 2012. Springer, Berlin, pp 367–385
7. Attrapadung N, Libert B, Peters T (2013) Efficient completely context-hiding quotable and linearly homomorphic signatures. In: Public-key cryptography–PKC 2013. Springer, Berlin, pp 386–404
8. Bernstein DJ, Buchmann J, Dahmen E (2009) Post-quantum cryptography. Springer, Berlin
9. Blake IF, Seroussi G, Smart N (1999) Elliptic curves in cryptography, vol 265. Cambridge University Press, Cambridge
10. Blumenthal D, Tavenner M (2010) The "meaningful use" regulation for electronic health records. N Engl J Med 363(6):501–504
11. Boneh D, Boyen X (2004) Short signatures without random oracles. In: Advances in cryptology–EUROCRYPT 2004. Springer, Berlin, pp 56–73
12. Boneh D, Boyen X (2008) Short signatures without random oracles and the SDH assumption in bilinear groups. J Cryptol 21(2):149–177
13. Boneh D, Boyen X (2011) Efficient selective identity-based encryption without random oracles. J Cryptol 24(4):659–693
14. Boneh D, Freeman DM (2011) Homomorphic signatures for polynomial functions. In: Advances in cryptology–EUROCRYPT 2011. Springer, Berlin, pp 149–168
15. Boneh D, Freeman DM (2011) Linearly homomorphic signatures over binary fields and new tools for lattice-based signatures. In: Public key cryptography–PKC 2011. Springer, Berlin, pp 1–16
16. Boneh D, Boyen X, Shacham H (2004) Short group signatures. In: Advances in cryptology–CRYPTO 2004. Springer, Berlin, pp 41–55

17. Boneh D, Gentry C, Lynn B, Shacham H (2003) Aggregate and verifiably encrypted signatures from bilinear maps. In: Advances in cryptology–EUROCRYPT 2003. Springer, Berlin, pp 416–432

18. Boneh D, Freeman D, Katz J, Waters B (2009) Signing a linear subspace: signature schemes for network coding. In: Public key cryptography–PKC 2009. Springer, Berlin, pp 68–87

19. Boyen X, Fan X, Shi E (2014) Adaptively secure fully homomorphic signatures based on lattices. IACR Cryptol 2014:916. http://eprint.iacr.org/2014/916 [ePrint Archive]

20. Catalano D (2014) Homomorphic signatures and message authentication codes. In: Security and cryptography for networks. Springer, Berlin, pp 514–519

21. Catalano D, Fiore D, Warinschi B (2011) Adaptive pseudo-free groups and applications. In: Advances in cryptology–EUROCRYPT 2011. Springer, Berlin, pp 207–223

22. Catalano D, Fiore D, Warinschi B (2012) Efficient network coding signatures in the standard model. In: Public key cryptography–PKC 2012. Springer, Berlin, pp 680–696

23. Catalano D, Fiore D, Warinschi B (2014) Homomorphic signatures with efficient verification for polynomial functions. In: Advances in cryptology–CRYPTO 2014. Springer, Berlin, pp 371–389

24. Charles D, Jain K, Lauter K (2009) Signatures for network coding. Int J Inf Coding Theory 1(1):3–14

25. Chaum D, Essex A, Carback R, Clark J, Popoveniuc S, Sherman A, Vora P (2008) Scantegrity: end-to-end voter-verifiable optical-scan voting. IEEE Secur Priv 6(3):40–46

26. Cheng C, Jiang T, Liu Y, Zhang M (2015) Security analysis of a homomorphic signature scheme for network coding. Secur Commun Netw 8(18):4053–4060 (2015). doi:http://dx.doi.org/10.1002/sec.1321

27. Coron J-S, Lepoint T, Tibouchi M (2015) New multilinear maps over the integers. Technical report, Cryptology ePrint Archive, Report 2015/162. http://eprint.iacr.org

28. Cortier V, Fuchsbauer G, Galindo D (2015) Beleniosrf: a strongly receipt-free electronic voting scheme. IACR Cryptology ePrint Archive, 2015:629

29. Cramer R, Shoup V (2000) Signature schemes based on the strong RSA assumption. ACM Trans Inf Syst Secur 3(3):161–185

30. Czap L, Vajda I (2010) Signatures for multisource network coding. Technical report, ArXiv

31. Dong J, Curtmola R, Nita-Rotaru C (2011) Practical defenses against pollution attacks in wireless network coding. ACM Trans Inf Syst Secur 14(1):7

32. Dutta R, Barua R, Sarkar P (2004) Pairing-based cryptographic protocols: a survey. IACR Cryptology ePrint Archive, 2004:64

33. El Gamal T (1984) A public key cryptosystem and a signature scheme based on discrete logarithms. In: Advances in cryptology, proceedings of CRYPTO'84, Santa Barbara, CA, August 19–22, 1984, proceedings, pp 10–18

34. Freeman DM (2012) Improved security for linearly homomorphic signatures: a generic framework. In: Public key cryptography–PKC 2012. Springer, Berlin, pp 697–714

35. Gennaro R, Halevi S, Rabin T (1999) Secure hash-and-sign signatures without the random oracle. In: Advances in cryptology–EUROCRYPT 1999. Springer, Berlin, pp 123–139

36. Gennaro R, Katz J, Krawczyk H, Rabin T (2010) Secure network coding over the integers. In: Public key cryptography–PKC 2010. Springer, Berlin, pp 142–160

37. Gentry C (2009) Fully homomorphic encryption using ideal lattices. In: Proceedings of the 41st annual ACM symposium on theory of computing, STOC 2009, Bethesda, MD, May 31–June 2, 2009, pp 169–178

38. Gentry C, Peikert C, Vaikuntanathan V (2008) Trapdoors for hard lattices and new cryptographic constructions. In: Proceedings of the fortieth annual ACM symposium on theory of computing. ACM, New York, pp 197–206

39. Gentry C, Sahai A, Waters B (2013) Homomorphic encryption from learning with errors: conceptually-simpler, asymptotically-faster, attribute-based. In: Advances in cryptology – CRYPTO 2013 – 33rd annual cryptology conference, Santa Barbara, CA, August 18–22, 2013. Proceedings, Part I, pp 75–92

40. Goldwasser S, Micali S, Rivest RL (1988) A digital signature scheme secure against adaptive chosen-message attacks. SIAM J Comput 17(2):281–308
41. Gorbunov S, Vaikuntanathan V, Wichs D (2015) Leveled fully homomorphic signatures from standard lattices. In: Proceedings of the forty-seventh annual ACM on symposium on theory of computing, STOC 2015, Portland, OR, June 14–17, 2015, pp 469–477
42. Guangjun L, Bin W (2013) Secure network coding against intra/inter-generation pollution attacks. Communications, China 10(8):100–110
43. Hiromasa R, Manabe Y, Okamoto T (2013) Homomorphic signatures for polynomial functions with shorter signatures. In: The 30th symposium on cryptography and information security, Kyoto
44. Hohenberger S, Waters B (2009) Short and stateless signatures from the RSA assumption. In: Advances in cryptology–CRYPTO 2009. Springer, Berlin, pp 654–670
45. Jing Z (2014) An efficient homomorphic aggregate signature scheme based on lattice. Math Probl Eng 2014:9 pp. Article ID 536527
46. Johnson R, Molnar D, Song DX, Wagner D (2002) Homomorphic signature schemes. In: Topics in cryptology - CT-RSA 2002, the cryptographer's track at the RSA conference, 2002, San Jose, CA, February 18–22, 2002, proceedings, pp 244–262
47. Kalra D, Ingram D (2006) Electronic health records. In: Information technology solutions for healthcare. Springer, Berlin, pp 135–181
48. Katz J (2010) Digital signatures. Springer, Berlin
49. Katz J, Waters B (2008) Compact signatures for network coding. http://www,cs.umd.edu/-jkatz/papers/NetworkC-dingSigs, pdf
50. Lee S-H, Gerla M, Krawczyk H, Lee K-W, Quaglia EA (2011) Performance evaluation of secure network coding using homomorphic signature. In: 2011 International symposium on network coding (NetCod). IEEE, New York, pp 1–6
51. Li F, Luo B (2012) Preserving data integrity for smart grid data aggregation. In: 2012 IEEE third international conference on smart grid communications (SmartGridComm). IEEE, New York, pp 366–371
52. Libert B, Peters T, Joye M, Yung M (2013) Linearly homomorphic structure-preserving signatures and their applications. In: Advances in cryptology–CRYPTO 2013. Springer, Berlin, pp 289–307
53. Libert B, Peters T, Joye M, Yung M (2015) Linearly homomorphic structure-preserving signatures and their applications. Des Codes Crypt 77(2–3):441–477
54. Makri E, Everts MH, de Hoogh S, Peter A, op den Akker H, Hartel PH, Jonker W (2013) Privacy-preserving verification of clinical research. In: Sicherheit 2014: Sicherheit, Schutz und Zuverlässigkeit, Beiträge der 7. Jahrestagung des Fachbereichs Sicherheit der Gesellschaft für Informatik e.V. (GI), 19–21 März 2014, Wien, Österreich, pp 481–500. http://subs.emis.de/LNI/Proceedings/Proceedings228/article4.html
55. Menezes AJ, Okamoto T, Vanstone S et al (1993) Reducing elliptic curve logarithms to logarithms in a finite field. IEEE Trans Inf Theory 39(5):1639–1646
56. Menezes AJ, Van Oorschot PC, Vanstone SA (1996) Handbook of applied cryptography. CRC Press, Boca Raton, FL
57. Molnar D (2003) Homomorphic signature schemes. PhD thesis, Citeseer
58. Paillier P (1999) Public-key cryptosystems based on composite degree residuosity classes. In: Advances in cryptology - EUROCRYPT'99, international conference on the theory and application of cryptographic techniques, Prague, May 2–6, 1999, proceeding, pp 223–238
59. Park C, Itoh K, Kurosawa K (1993) Efficient anonymous channel and all/nothing election scheme. In: Advances in cryptology - EUROCRYPT'93, workshop on the theory and application of cryptographic techniques, Lofthus, May 23–27, 1993, proceedings, pp 248–259
60. Rivest RL, Shamir A, Adleman L (1978) A method for obtaining digital signatures and public-key cryptosystems. Commun ACM 21(2):120–126

61. Robertson A, Cresswell K, Takian A, Petrakaki D, Crowe S, Cornford T, Barber N, Avery A, Fernando B, Jacklin A et al (2010) Implementation and adoption of nationwide electronic health records in secondary care in England: qualitative analysis of interim results from a prospective national evaluation. BMJ 341:c4564
62. Shao J, Zhang J, Ling Y, Ji M, Wei G, Ying B (2013) Multiple sources network coding signature in the standard model. In: Internet and distributed computing systems. Springer, Berlin, pp 195–208
63. Shor PW (1994) Algorithms for quantum computation: discrete logarithms and factoring. In: 35th annual symposium on foundations of computer science, 1994 proceedings. IEEE, New York, pp 124–134
64. Wang Y (2010) Insecure "provably secure network coding" and homomorphic authentication schemes for network coding. IACR Cryptology ePrint Archive, 2010:60
65. Wang F, Hu Y, Wang B (2013) Lattice-based linearly homomorphic signature scheme over binary field. Sci China Inf Sci 56(11):1–9
66. Wang F, Wang K, Li B, Gao Y (2015) Leveled strongly-unforgeable identity-based fully homomorphic signatures. In: Information security. Springer, Berlin, pp 42–60
67. Waters B (2005) Efficient identity-based encryption without random oracles. In: Advances in cryptology–EUROCRYPT 2005. Springer, Berlin, pp 114–127
68. Xia Z, Culnane C, Heather J, Jonker H, Ryan PY, Schneider S, Srinivasan S (2010) Versatile prêt à voter: handling multiple election methods with a unified interface. In: Progress in cryptology-INDOCRYPT 2010. Springer, Berlin, pp 98–114
69. Yan W, Yang M, Li L, Fang H (2012) Short signature scheme for multi-source network coding. Comput Commun 35(3):344–351
70. Yang L, Li F (2013) Detecting false data injection in smart grid in-network aggregation. In: 2013 IEEE international conference on smart grid communications (SmartGridComm). IEEE, New York, pp 408–413
71. Yu Z, Wei Y, Ramkumar B, Guan Y (2008) An efficient signature-based scheme for securing network coding against pollution attacks. In: INFOCOM 2008. The 27th conference on computer communications. IEEE. IEEE, New York
72. Yun A, Cheon JH, Kim Y (2010) On homomorphic signatures for network coding. IEEE Trans Comput (9):1295–1296
73. Zhang N (2010) Signatures for network coding
74. Zhang P, Yu J, Wang T (2012) A homomorphic aggregate signature scheme based on lattice. Chin J Electron 21(4):701–704
75. Zhang J, Shao J, Ling Y, Ji M, Wei G, Ying B (2015) Efficient multiple sources network coding signature in the standard model. Concurr Comput Pract Exp 27(10):2616–2636. doi:http://dx.doi.org/10.1002/cpe.3322
76. Zhao F, Kalker T, Médard M, Han KJ (2007) Signatures for content distribution with network coding. In: IEEE international symposium on information theory, 2007. ISIT 2007. IEEE, New York, pp 556–560

Printed in the United States
By Bookmasters